More Praise for KINGDOM OF PLAY

"This engaging survey of animal play—from bees to humans—not only reveals the astonishing array of play behaviors but the motivations behind such escapades as octopuses tossing balls and crows devising their own snow sleds. Toomey also searches for an overarching explanation for animals' playful inclinations and finds it in natural selection, Darwin's key driver of evolution. Overall, a fun and thought-provoking examination of why we and all animals play."

> —Virginia Morell, *New York Times* bestselling author of *Animal Wise: How We Know Animals Think and Feel*

"In nearly forty years of studying the play of animals, I have witnessed an extraordinary growth in knowledge—quite simply, we know so much more now than when I began. This marvelous book makes what we now know accessible to a wide audience, showing how play has grown from a frivolous backwater to a central focus of research. But insiders can be fooled into thinking they know more than they do, and Toomey's book is not shy in pointing out that there are fundamental mysteries that remain to be explained. Indeed, the book takes us on a historical journey that places our successes and our continuing shortfalls in context. Play remains an enigma, but an enigma that deserves our attention."

> —Sergio M. Pellis, professor of animal behavior and behavioral neuroscience, University of Lethbridge, and coauthor (with Vivien Pellis) of *The Playful Brain: Venturing to the Limits of Neuroscience*

"David Toomey has a knack for zeroing-in on scientific conundrums, digesting the studies and research about them, and then explaining them in accessible and often humorous prose. Surveying theories about animal play from the nineteenth century to the present, he forwards bold and surprising hypotheses about the nature and evolution of play. And as we learn about animals who play with humans, and people who play with and as nonhuman animals, the implications of this study encompass the entire living kingdom. A brilliant, instructive book!"

> —Randall Knoper, professor emeritus of English and author of *Literary Neurophysiology: Memory, Race, Sex, and Representation in U.S. Writing, 1860–1914*

KINGDOM OF PLAY

What Ball-Bouncing Octopuses,
Belly-Flopping Monkeys, and Mud-Sliding
Elephants Reveal about Life Itself

DAVID TOOMEY

SCRIBNER

New York   London   Toronto   Sydney   New Delhi

Scribner
An Imprint of Simon & Schuster, LLC
1230 Avenue of the Americas
New York, NY 10020

First Scribner hardcover edition March 2024

SCRIBNER and design are registered trademarks of Simon & Schuster, LLC.

Simon & Schuster: Celebrating 100 Years of Publishing in 2024

For information about special discounts for bulk purchases,
please contact Simon & Schuster Special Sales at 1-866-506-1949
or business@simonandschuster.com.

The Simon & Schuster Speakers Bureau can bring authors to
your live event. For more information or to book an event,
contact the Simon & Schuster Speakers Bureau at 1-866-248-3049
or visit our website at www.simonspeakers.com.

Interior design by Kyle Kabel

Manufactured in the United States of America

1   3   5   7   9   10   8   6   4   2

Library of Congress Cataloging-in-Publication Data has been applied for.

ISBN 978-1-9821-5446-2
ISBN 978-1-9821-5448-6 (ebook)

Contents

Introduction ix

CHAPTER 1 Ball-Bouncing Octopuses: What Is Play? 1

CHAPTER 2 The Kalahari Meerkat Project:
The Hypotheses of Play 17

CHAPTER 3 Tumbling Piglets and Somersaulting Monkeys:
Training for the Unexpected 41

CHAPTER 4 "Let's Go Tickle Some Rats":
The Neuroscience of Play 65

CHAPTER 5 Courtly Canines: Competing to Cooperate
and Cooperating to Compete 93

CHAPTER 6 Wood Thrush Songs, Herring Gull Drop-Catching,
and Bowerbird Art: Play as the Roots of Culture 113

CONTENTS

CHAPTER 7   Memes and Dreams: Dreaming
as Playing without a Body 137

CHAPTER 8   The Evolution of Play 153

CHAPTER 9   Innovative Gorillas: The Surprising
Role of Play in Natural Selection 183

CHAPTER 10   Playing Animal 205

EPILOGUE   Play, Life, and Everything 233

Acknowledgments 239

Notes 241

Bibliography 253

Index 267

Introduction

In the winter of 2020–21, many of us were anxious, lost, and alone. We were talking to screens and tired of talking to screens. It was difficult to imagine a future that would be anything other than a long succession of days. But one morning in January, the Smithsonian National Zoo in Washington, DC, posted videos from its Panda Cams. Five inches of snow had fallen overnight, and the zoo's two adult giant pandas, Mei Xiang and Tian Tian, were playing in it—rolling, somersaulting, and sliding slowly down a long, curving slope. The video was shared widely. Friends sent it to friends, grandchildren to grandparents, dog lovers to cat lovers. Many who watched it set aside the hardships of that terrible year and for a moment felt a surge of happiness. The pandas' play came as a reassurance, a reminder not of mere normalcy (and there was much talk that year of a return to normalcy), but welcome evidence that for all its woes, the world still held a place for exuberance, even joy. As to their manner of play—most particularly one panda sliding on his back headfirst—it was a willing surrender to gravity and momentum, a trust that all would be well in the end. In that way, it was even an act of faith when we needed most to see one.

INTRODUCTION

* * *

Animals at play inspire wonder, delight, and even awe. Yet until recently, scientists have given animal play little attention. It's a curious oversight. Play in humans, especially human children, has been a subfield of psychology for more than a century. Other human behaviors—mating, grooming, and caring for young, for instance—have been illuminated by research about the same or similar behaviors in other species. We might expect that library shelves would be creaking under the weight of books, doctoral theses, and journal articles describing and theorizing about animal play. But they aren't. By comparison to other kinds of animal behavior, the number of studies on animal play is meager. There is no journal of animal play, no handbook or encyclopedia of animal play, no institute of animal play, and no college or university with an academic department dedicated to its study. And in more than 120 years, only five books have treated the topic exclusively.

Why has this subject been so neglected?

Likely for several reasons. For one, play is hard to define. Practitioners in fields ranging from child psychology to cultural anthropology have offered definitions ranging from "play underlies all creativity and innovation," to "play is cruel sport, teasing and competition," to "play is the source of rituals and myths by which we structure our lives."[1] Play can be difficult to distinguish from other behaviors—say, exploring or mating. Even when play *is* well-defined and well-identified, it can be difficult to observe since most animals that play do so for only a few minutes a day. Another reason scientists neglect play is less about what behavior in animals counts as play and more about what behavior in scientists counts as *work*. Until recently, many of the committees and granting agencies that approve and fund scientific research considered animal play undeserving of serious inquiry. No one knew this better than the late Jaak Panksepp, a pioneer researcher of animal emotions. In 1990 he told an interviewer that play is "a topic that many still deem as relatively frivolous and unimportant."[2] Ethologist Gordon Burghardt notes, with evident chagrin, that when he mentions his work in animal play to other scientists, their response is often only "amused interest and a shared story about a pet."[3] Even those who've conducted research in the behavior have suggested that efforts to understand it may prove futile. Philosopher Drew Hyland, in his 1984 work *The Question of Play*, doubts that play can be rigorously defined, let alone analyzed. Robert Fagen, in his masterly *Animal Play Behavior*, after spending nearly five hundred pages surveying the research and thinking surrounding the subject, called play "a pure aesthetic that frankly defies science."[4]

These sentiments have had practical consequences. Scientists who think foundations and academic committees will not fund research into animal play do not design their research agendas around it. Those same foundations and academic committees receive no applications for research into animal play and therefore assume that it is of little interest to scientists and so not worth funding and thus offer fewer grants in the area. Scientists who oversee graduate students who expect to be funded advise them to look elsewhere for a dissertation topic. In time, those graduate students gain positions at colleges and universities and impart the same advice to their own students for the same reasons. And so on to the next generation of scientists, and the next. In 1980 eminent naturalist, entomologist, and author E.O. Wilson summarized the challenges of studying animal play, writing, "No behavioral concept has proven more ill-defined, elusive, controversial, and even unfashionable."[5]

But things are changing. In recent years, the study of animal play has been given renewed impetus by two developing areas of research. One area is animal culture. Culture and play are intertwined, and any understanding of animal culture is likely to be facilitated by an understanding of animal play. Another area is neuroscience. New techniques and technologies for brain imaging (especially positron-emission tomography and magnetic resonance imaging) are leading to ever more finely detailed mapping of the neural networks. In time, they may show how play changes the brain's chemistry and neural pathways, and, alternately, how that chemistry and those neural pathways enable play.

A Bundle of Mysteries

Young adult ravens dive and turn by tucking in one wing, spreading the wing again, and turning back. They roll over midflight and chase one another, "dive-bombing" and feinting in and out of one another's way. A bottlenose dolphin witnessed its companions being taught to tail walk for a public aquarium performance and, after being released into the wild, began to tail-walk unprompted; to the astonishment of researchers, its wild companions began to tail-walk, too. Elephants have been seen sliding down muddy slopes, some on bellies, some on backsides. The ravens and dolphins might be said to be embellishing ordinary activities, developing reflexes and skills, or engaging in courtship behavior. But mud sliding would seem to be utterly irrelevant to any imaginable elephantine need.

All these behaviors are play, and they present a problem to the scientists who study animal behavior, known as ethologists. Since these behaviors take time and energy, and since they can be dangerous, most ethologists assume that play must help an animal to survive or reproduce, and that it must have one or more adaptive advantages precisely because it has so many obvious *dis*advantages. Yet they have not agreed upon what those advantages might be.

The question of why animals play invites a great many more. The subject might seem not so much a coherent field of study as a loose bundle of mysteries. There are questions of taxonomy. Which animals play? Which don't? There are problems of definition and identification. What exactly is play? How can we be sure a certain behavior is play and not, say, exploration? There are questions about the role of play in an animal's development. Do the ones that play do so at particular stages in their lives? There are questions of heredity and environment. To what degree is play instinctive? To what degree is it learned? There are questions of play's relation to an animal's brain and nervous system. What neural mechanism or process enables play or brings it about? Are certain parts of the brain necessary? Are certain *kinds* of brains necessary? Then there are questions of evolution and natural selection. When, during the long history of life on earth, did play first appear? Exactly how did it evolve? And what of the future? Is it possible that in animal play we are seeing the beginnings of animal culture?

Animal play then is not just another scientific mystery—it's a collection of them, and different from most. The phenomena at the core of many scientific mysteries—quantum entanglement and dark matter, to name two—are rather distant from our everyday lives. To study them requires specialized knowledge and perhaps large and expensive instruments. Animal play is all around us, however, a thing we see every day. We don't need an advanced degree or a particle accelerator to study it. We need only to watch an animal and give it our attention.

And there is a reason it's worth that attention.

The Characteristics of Play Are the Characteristics of Natural Selection

Since its presentation 160 years ago, Darwin's theory of evolution by natural selection has been much refined and elaborated upon. In the twentieth century, Mendelian genetics explained its mechanism; the theory has been amended with findings from microbiology, developmental biology, and most recently epigenetics. For all this, at its core the theory is unchanged. Natural selection is a filter, or a series of filters, straining out detrimental variations and allowing advantageous ones to pass through so that with each generation an organism becomes better adapted, or "more fit."

Natural selection possesses a number of specific and well-defined characteristics. It is, for instance, purposeless. It has no intention, and no objective,* and as Darwin averred, it "includes no necessary and universal law of advancement or development."[6]

---

* Since no being or entity is choosing or selecting, "natural selection" is not a selection in the usual sense. Darwin was unsatisfied with his nomenclature, writing, "I suppose natural selection was a bad term, but to change it now, I think, would make confusion more confounded. Nor can I think of better." Letter to Charles Lyell, June 6, 1860, Darwin Correspondence Project. Incidentally, *survival of the fittest* was the coinage of English biologist and sociologist Herbert Spencer. Darwin used the phrase in later editions of *On the Origin of Species.*

It is provisional. The evolution of any organism is a response to whatever conditions are present at a given place and moment. It is open-ended. The evolution of any organism has no moment of arrival and no end point—a fact highlighted in the final paragraph of *On the Origin of Species*, a slowly building crescendo whose final note hangs in the air and never quite resolves: the forms of life, it concludes, are even now "being evolved." In all these ways, natural selection is like play. As we'll see, there are other similarities—so many that if you could distill the processes of natural selection into a single behavior, that behavior would be play. Alternately, if you were to choose an evolutionary theory or view of nature for which play might seem to be a model, it would be natural selection.

Natural selection is not merely an important activity in which living organisms participate. It is the essential activity, the activity that distinguishes them from everything else. Organisms do many things: they grow, convert matter to energy, and eventually fade out of existence. But such activities are not exclusive to organisms. Candle flames and stars also do these things, and they are not living. One might note that organisms do something candle flames and stars cannot do: they reproduce themselves. But crystals reproduce themselves, and they are not living. The one thing living organisms do that flames, stars, and crystals don't do—and can't do—is evolve by natural selection.

Since life is best defined as that which evolves by natural selection, and since natural selection shares a great many features with play, we require no great leap of reasoning to arrive at the thesis presented and developed in the chapters that follow. Life itself, in the most fundamental sense, is playful.

KINGDOM OF PLAY

CHAPTER 1

Ball-Bouncing Octopuses: What Is Play?

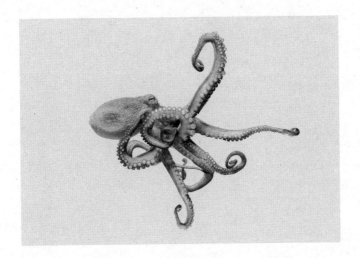

One day in 1997, psychology professor Jennifer Mather answered her phone to hear the excited voice of Roland Anderson, her collaborator in a rather unusual study of animal behavior. "She's bouncing the ball!" He was speaking in figurative terms, but ones he knew Mather would understand. The "she" was an octopus perhaps two or three years old, swimming in a tank at the Seattle Aquarium.* The "ball" was an Extra Strength Tylenol bottle weighted to float just beneath the surface. And the "bouncing" wasn't exactly bouncing. Octopuses have an exhalant funnel, a siphon near the side of the head, through which they can jet water. The octopus had held the bottle with her arms and let it go; she then aimed her funnel at it and released a jet of water in its direction, sending it to the other end of the aquarium, where the water flow returned it to her.† She was doing it again and again. After Anderson had seen her perform the feat sixteen times, he decided it was time to call Mather.

* The subject compels us to make certain choices in language and phrasing. I'll use gender pronouns when referencing an account or study of a specific animal or animals whose sex is specified, as is the case here. When an animal's or the animals' sex is undetermined or unimportant, I'll use it or them. *Homo sapiens* are animals, yet it will often be convenient to treat them as separate from others. As for those others—the phrase *nonhuman animal* references millions of species as though they are a single category and so fails to recognize those species' individuality—an especially significant oversight here, since much play is initiated by individual animals. Moreover, the phrase is awkward; its repeated use would hinder reading. So, with reservations and the hope that you will recall that those more than a million species are as different from one another as they are from us, I'll bow to precedent and use the word *animal* to mean the nonhuman sorts.

† The arms of an octopus have suction cups along their entire length, while tentacles by definition have suction cups only at their outer extremities. Incidentally, the standard plural of *octopus* is *octopuses*, and since the word *octopus* is derived from Greek, the Greek plural form *octopodes*, while uncommon, is also correct.

Jennifer Mather is a preeminent researcher on octopus behavior, and a peripatetic academic, having done research in many places. During a dive in the waters off Bermuda, she noticed an interesting behavior of the common Atlantic octopus known as *Octopus vulgaris*. It found a small rock, tucked it under its arms, and placed it near its den. The octopus then retrieved another rock. And another and another. It slid into the den and pulled all the rocks behind it, effectively barring the entrance from intruders before going to sleep. Octopuses had been known to demonstrate rudimentary intelligence, but Mather recognized this as something far more sophisticated. Since the octopus was not reacting to any existing threat, but rather in anticipation of one, it had demonstrated foresight and planning.[1]

Like Mather, Roland Anderson was a scuba diver. His preferred environs were the rather colder waters of Puget Sound, where he was known to dive even at night and in the rain, once just to look for small *Octopus rubescens* that had found homes in beer bottles resting on the ocean floor. Anderson's vocation and avocation were happily one. Professionally, he was a marine biologist employed at the Seattle Aquarium, where his specialties included the natural history and behavior of octopuses. While performing his rounds one morning, Anderson discovered that one of the aquarium's giant Pacific octopuses had dug up the gravel from the bottom of its tank, bitten through the nylon cable ties that attached the filter to the tank, and torn the filter into small pieces. Anderson did not understand the octopus's motivation for destroying the filter, but he did know that its actions, like those of Mather's octopus, were methodical, requiring both foresight and planning.

Mather and Anderson first met at a conference and found they shared a fascination for a certain order of cephalopod mollusk as well as a suspicion that there was more going on in their minds than many believed. They wondered whether octopuses played. The creatures had a penchant for manipulating objects, which indicated both intelligence and curiosity. The step from manipulating objects to playing with them might seem small, and in the terms of behavior, it would be. But in terms of taxonomy, it would be a giant step—or rather a leap—across whole phyla. Mammals and birds, long recognized as playful, are members of the phylum Chordata. But octopuses are cephalopods, of the phylum Mollusca.

Early in the third century CE, Roman natural historian Claudius Aelianus wrote, "Mischief and craft are plainly seen to be the characteristics of this creature."[2] In the years since, they've become no less mischievous and crafty, as attested by researchers whose instruments they are known to take apart. When marine biologist Jean Boal and her colleagues noted octopuses' "relative intractability as experimental subjects,"[3] they might have been speaking for many a weary cephalopod researcher. Octopus behavior certainly seems playful. Yet no one had attempted to demonstrate that it *was* play or to prove, under experimental controls, that octopuses played. Real, empirical evidence of octopus play would be a finding of some consequence. It would mean that the behavior had evolved in two phyla that had diverged 670 million years ago.[4]

In the 1990s many experiments gauged octopus intelligence and curiosity. Most had been of the rudimentary, stimulus-response sort. An object, say a mussel shell, would be dropped into an octopus's tank. An octopus seeking food would explore the shell, use its arms to push, prod, and turn it over. When it found no food and nothing more worth investigating, the octopus would lose interest in the shell and leave it alone.

If the same experiment was performed with an especially curious octopus, the response might be different. When the shell first appeared, the octopus would explore it. The feel of striations on the shell's surface might stimulate further exploration, and eventually the octopus would find the shell's inner concavity. Its smooth surface, an especially intriguing contrast to the striations, might stimulate more study. So it would go, a recurring cycle of stimulus and investigation, with each stimulus prompting further investigation and each investigation uncovering another stimulus. The cycle would greatly extend the period of investigation, with a curious octopus taking far longer to explore the shell before losing interest.

Mather and Anderson thought that in that slide from exploration to habituation some behavior might justifiably be called play. To distinguish play from exploration would be difficult, but they designed an experiment that they hoped might accomplish just that. They chose as their test subject the giant Pacific octopus (*Enteroctopus dofleini*), a species well adapted to the cold, oxygen-rich water of the coastal North Pacific. It's the largest octopus species; an adult may weigh more than 110 pounds and live for four or five years. The individuals to be tested, five males and three females, were somewhat lighter and younger—"subadults" two to three years old that weighed between two and twenty pounds.

Mather and Anderson knew that they might prompt play by giving the octopuses something interesting to investigate. They decided upon four plastic pill bottles weighted to just above neutral buoyancy. Octopuses have limited color vision but good vision for light intensity, so two bottles were painted white and two painted black. And since octopuses are sensitive to texture, one white and one black bottle were given smooth surfaces, the other two rough ones. Each octopus was allowed ten opportunities or "trials" lasting thirty minutes, which Mather and Anderson deemed sufficient time for the octopuses either to engage with the bottles or to demonstrate no interest in them.

All eight octopuses made contact with the bottles in some fashion, either by pressing a sucker against a bottle, curling an arm around it, or using an arm to press the bottle against its mouth. Mather and Anderson judged these actions to be investigation. Two octopuses, though, engaged with the bottles in a manner that looked like play. Octopus 8 used its funnel to create a jet of water that pushed a bottle to the wall of the aquarium and back. Octopus 7 issued a water jet that caused the bottle to travel "in a circular path . . . around the aquarium periphery," prompting Anderson's phone call.[5]

Octopuses had been known to use exhalant funnels to propel themselves, to clean detritus at the mouth of their dens, and to push away irritations such as scavenging fish and stinging sea anemones. In captivity, octopuses often use the funnels to repel a food they regard with some contempt—say frozen shrimp. But as far as anyone knew, they had never been used like this. Octopuses 7 and 8 were not pushing the bottle away; on the contrary, they had found an interesting way to make it return.

Mather has addressed all manner of audiences—from professional animal behaviorists to schoolchildren—hundreds of times, but she speaks with such deliberation that you might think that as she voices each word, she considers it anew. When she later adopted Anderson's metaphor, she was careful to explain it. "That's just exactly the kind of thing we do when we bounce a ball," she told an interviewer. "When you bounce a ball, you are not trying to get rid of the ball, you are trying to figure out what you can do with the ball."[6]

The Categories of Play

Ethologists have established three categories of play. There is *solitary play*, like the frolic of a pony alone in a field. There is *social play*, the play wrestling of juvenile chimpanzees. And there is *object play*, the stick chasing and retrieval of a puppy—or an octopus using its exhalant funnel to propel a pill bottle. These categories are distinct, and for ethologists studying a certain kind of play in a particular animal, that focus has advantages. But the price paid for it is the neglect of a more complicated reality. Animals, innocent of ethologists' categories, mix kinds of play. Two puppies wrestling a stick from each other may be indulging in both object and social play. Some play may be combined with behavior that is not play at all: a crow manipulating a twig might be engaging in both object play and exploration, and lorises in social play might also be practicing courtship behavior. Adding to the difficulty of defining play is that solitary, object, and social play are not all the play there is. Psychologists have been studying play in human children for more than a century and have identified many other types of play—parallel play, pretend play, play mothering, and construction play among them. That these are present in animals as well is increasingly evident.

This multiplicity of play is part of its allure, and for ethologists focusing on a particular animal, it's of little concern. But for those who wish to gain a comprehensive understanding of play that includes its origins and its evolution, that multiplicity presents a problem. They need to compare play across species and across whole classes of animal. To do that, they need to know exactly what they are comparing, and to be assured that they are comparing not merely what looks like the same behavior, but what *is* the same behavior. For that they need a general, all-purpose definition of play, one that is freestanding and unambiguous, yet accommodates the behavior in all its varieties and permutations. And if we are to undertake a thorough study of play, we'll need one, too.

Defining Play

The editors of the *New Oxford American Dictionary* list ten definitions of *play*, operating in most or all realms of human and animal endeavor: six verbs with variations, four nouns with variations, and scores of compound words and phrases. There is, it seems, a lot of "play" in the word *play*.

Definitions of play are admirably wide-ranging, but most have limitations. Some use tautologies, defining it, for instance, as an activity that's fun. Whatever fun is, it's at least as difficult to define as play. Some definitions rely upon terms such as *pleasure* or *satisfaction*—terms perhaps more precise than *fun*, but difficult to gauge objectively. Other definitions presuppose a function for play, when no single function or set of functions has been identified. Still others suggest that it's a behavior with *no* function[7] and so do not distinguish play from other behaviors with no clear function, among them the repetitious and compulsive movements of caged or distressed animals. Moreover, the claim that play has no function is not entirely true. Much play, for instance, requires exercise, and exercise benefits an animal by increasing the oxygen-carrying capacity of the blood.

Some definitions list criteria for play and qualify those criteria with words such as *may be* or *might include.* There's this: "Play is all motor activity performed postnatally that appears to be purposeless in which motor patterns from other contexts may often be used in modified form and altered temporal sequencing."[8] Obviously if we want something definitive in our definition, equivocating phrases such as *may often be used* are of little help.

Playing animals are seldom able and willing to describe their experience to researchers, yet many definitions assume knowledge of that experience. One definition of play, a behavior "performed for its own sake," ascribes intention. Another, in calling play "pleasurable," ascribes emotion. Still others assign the player a motive or—as in "behavior performed without the 'serious point' that such behavior has in its normal context"[9]—a lack of one.[*]

In theory, one might define an emotion such as happiness by describing its neurological roots. Perhaps play is likewise the

---

[*] In *Animal Play Behavior* Robert Fagen proffers a definition that sidesteps most of these failings. It's a mouthful, including a dash of somewhat specialized language, but it's admirably comprehensive. It's also provisional. "For purposes of this book," he writes (p. 21), "play includes nonagonistic fighting and chasing maintained by social cooperation; solo solitary and rotational movements performed in the absence of threatening predators, parasites, and conspecifics; developing solitary or manipulative behavior repeated with slight variation at a previously established level of mastery; and diversive effector interactions with an inanimate object subsequent to the termination of an initial phase of sensory and mastery activity, including exploratory manipulation, directed toward the object."

activity of a specific neural mechanism. Studies of the brains and nervous systems of rats and mice have made some progress in that direction. But since such studies are undertaken in laboratories, their subjects may behave differently from how they would in their natural environments, rendering the results suspect. Moreover, since many such studies require autopsies and "sacrificing" the subject animal, they cannot humanely be undertaken with most animals. Many would say with *any* animals. And even if a researcher could capture an animal's inner experience by monitoring the activity of its neural circuitry without harming it and as it played in its natural environment, that would capture only the inner experience of one animal representing only a single species among the great many that play.

We don't know what an animal experiences as it plays. To define play, then, we must for now limit our definition to what it looks like. But that brings its own set of problems.

Identifying Play

We might assume that two dogs tussling, or a cat swatting a toy mouse, or otters cavorting and splashing in a river, are playing, and we might be right. Their play looks a lot like our play, and since they are mammals like us, perhaps it *is* like our play. But in many animals, the behaviors of play, courtship, and fighting intermingle and are not always easy for a researcher to distinguish.

This difficulty is compounded by two behaviors that are not play, but closely resemble it.

The first is what ethologists call *stereotypies*, which one behavioral biologist defines as "a behavior pattern that is repetitive, invariant and has no obvious goal or function."[10] While play indicates well-being, stereotypies signal that the animal is deprived of stimulation or feeling stress. We've all seen it: a parrot repeatedly jumping from place to place in a cage, a wolf in an enclosure pacing the same course for hours, or you or me chewing a pencil or twirling a strand of hair. The other behavior difficult to discriminate from play is *exploration*—that is, moving through an unfamiliar area or manipulating an object to learn about it.

Play is not all that easy to recognize even in our fellow *Homo sapiens*. Humans at play often exhibit signs of pleasure—a laugh, a shout, a fist pump in the air. But not always. Consider the taut expression of a chess player planning a move or a soccer player advancing a ball toward the goal. It has a name, *game face*, defined by the *New Oxford American Dictionary* as "a sports player's neutral or serious facial expression, displaying determination and concentration." Few would associate this demeanor with pleasure, yet those wearing it are, by definition, playing. We cannot readily identify pleasure in members of our own species, with whom we share physiognomy. How much more difficult is it then to identify pleasure in animals somewhat more distant from us? How can we know what play looks like in a crocodile? An Atlantic salmon? A termite?

Several ethologists have given thought to these questions, but probably none more deeply or carefully than Gordon Burghardt. Burghardt is bespectacled and bearded and has the soft-spoken, meticulous, and thorough manner of a country lawyer. His fellow scholars allow that a great range of animal species play, most of them mammals and birds. But Burghardt finds evidence for play—if not actual proof—in animals many would say are incapable of it: a saltwater crocodile, cichlid fish, even honeybees. Some might call such claims a bit outlandish, and coming from someone else they might well be. Burghardt, though, is an Alumni Distinguished Professor in the Departments of Psychology and Ecology & Evolutionary Biology at the University of Tennessee, Knoxville. His career is as distinguished as it is long. If his ideas are unorthodox, he makes them credible because he arrives at them by careful, measured, and well-reasoned steps. One of Burghardt's colleagues called him a "sensible radical."[11]

Burghardt found the definitions of play put forth by many researchers lacking. He wanted a general, all-purpose definition of animal play that would cover solitary, object, and social play and their permutations, and that could also be used to identify play in the behavior of animals that, most assume, don't play. A definition of play that would meet this standard would avoid tautologies and ambiguous terms. It would not presuppose an adaptive advantage or advantages. It would eschew equivocating phrases such as *may be* and *might be*. It would not assume the mental or emotional state of the player. And it would clearly differentiate play from stereotypies, exploration, and other behaviors that it might resemble. In his 2005 book *The Genesis of Animal Play* Burghardt maintained that for behavior to count as play, it must have five characteristics.

First, play must be "nonfunctional"—that is, it must not in any obvious way serve an animal's need to survive or reproduce.

Second, play must be purely voluntary and not a forced response to some external influence.

Third, play must be obviously unlike the animal's other behaviors.

Fourth, the movements of an animal at play must be repeated, thus distinguishing them from the actions of an animal exploring, which are varied. But those same movements must be repeated in a different order and with modifications, thus distinguishing them from the movements of stereotyped behavior, which is unvaried.

Fifth and finally, Burghardt maintains that play occurs when and only when the animal is well-fed, safe, and healthy, and when no external influence is forcing its behavior. This is a check on the second criterion, as some might object that its use of the word *voluntary* presumes the inner experience of the animal—in this case, intention—when we cannot know what that experience might be. The absence of any external influence likely means that the animal is not playing because it has to; it is playing because it wants to.

If you see an animal behaving in a way that is nonfunctional, voluntary, and characterized by repeated but varied movements, and if that animal is well-fed, safe, and healthy, Professor Burghardt says you are seeing an animal at play.

Let's apply Burghardt's criteria to the behavior of Octopuses 7 and 8. Since the octopuses found the bottles unsuitable as food and even less suitable as mates, their behavior was nonfunctional, unrelated to survival or reproduction. Since nothing compelled the octopuses to engage with the bottles, their behavior was voluntary. Might it have been exploration? Likely not. Exploration and play look similar, especially when both involve objects, but there's a crucial difference. As Mather noted, an animal exploring an object is gathering information about it; an animal playing with an object is discovering what can be done with it. The octopuses were clearly doing the latter. The aquarium's water flow returned the bottles to the general vicinity of the octopuses, but in slightly different places each time. To continue to engage the bottles, the octopuses adjusted the direction of their water jets and repositioned themselves. Thus the behavior was characterized by varied movements. Finally, an animal exploring a new environment is on the lookout for two things: resources and threats—to put it roughly, things it can eat and things that can eat it. The possible presence of the latter means that an animal exploring is not entirely at ease. Only after it is satisfied that no danger is nearby does it relax, and only when it is relaxed can it move on to other activities—such as play. The octopuses were well-fed, healthy, and—alone in a tank of water with no external influences—safe.

The "ball bouncing" of Octopuses 7 and 8 met each of Burghardt's criteria. But some researchers, upon learning of Mather and Anderson's experiment, were skeptical. Jean Boal said the behavior "could reflect boredom, like a cat pacing."[12] Burghardt himself countered that because a behavior is a response to boredom does not mean it isn't play, and that boredom might actually trigger play.[13]

In 2003 Mather and several colleagues performed a follow-up.[14] The prospective players this time were seven *Octopus vulgaris*, and the prospective playthings were LEGO pieces and floating bottles on strings. The range of responses was a reminder of octopuses' individuality, a characteristic familiar to aquarium attendants. Some ignored the objects altogether. Others made contact when the objects were first introduced and ignored them thereafter. Two octopuses brought the objects into their dens. Only one octopus exhibited behavior the researchers defined as "full play," but others engaged in what they called "play-like interactions"—for instance, pushing and pulling a LEGO piece, towing a bottle, and passing a LEGO piece from one arm to another. The experiment, Mather and colleagues concluded, found "promising evidence that *O. vulgaris* exhibits play behavior."[15]

Play and Natural Selection

Let's highlight three aspects of the behavior of Octopuses 7 and 8 that also characterize natural selection. It is purposeless. The octopuses using their exhalant funnels to propel a bottle have no particular goal, no plan or agenda. They are merely moving the bottle and mildly interested in seeing what happens when it's moved. Their behavior is provisional. The octopuses may stop propelling the bottle in a certain direction upon finding a more interesting result by propelling it in a different direction. Their behavior is ongoing and open-ended. It may reach a certain stage—say, the bottle completing a circuit of the tank—but the activity doesn't end at that moment; nor is it a reason for the octopuses to end it.

Octopuses 7 and 8, when they propelled the bottles back and forth and around the aquarium, embodied three features of Darwin's theory.

CHAPTER 2

The Kalahari Meerkat Project:
The Hypotheses of Play

Shortly after sunrise in the austral summer of 1999, deep in South Africa's Kalahari Desert, doctoral student Lynda Sharpe was crawling through sour grass and braving swarms of stinging insects. Her attention was focused on a group of meerkats, a species of southern Africa mongoose, which were asleep in the sun outside their den. Moving quickly and quietly, she caught one, snipped a bit of fur, and marked her tail with a felt-tip pen. Then she did the same to another, and another.

Sharpe was in her early thirties, with open features and freckled skin that made her seem still younger. She had a ready sense of humor, often joking that the study of South African fauna was a peculiar career choice, and called the meerkats, sharp-clawed predators who occasionally kill one another's young, "my little darlings." But her intent that morning was quite serious. She had marked the meerkats so she'd be able to tell them apart and learn who among them was playing more, and with whom. If she knew these things, she hoped to derive an answer to a long-standing question in animal behavior: *Why* do animals play?

The Costs of Play

Play seems to have little to do with survival in any obvious way. Quite often it seems to *hinder* survival. Play is costly, taking energy and time that might be better spent hunting, foraging, or mating; it can also be dangerous, leading to injury and death. Siberian ibex kids that play on cliffs occasionally fall,[1] and juvenile vervet monkeys prefer to play at some distance from adults, making them more vulnerable to predation by yellow baboons.[2] One of the more harrowing accounts of play's costs comes from Robert Harcourt, a zoologist now at Macquarie University in Sydney, Australia. He observed South American fur seals on a beach at Punta San Juan, where about two thousand pups are born every year. From January to October 1988, Harcourt observed the seals at all hours of daylight; on 102 occasions he saw pups attacked by southern sea lions, with twenty-six pups killed. Of these, twenty-two were playing in the shallow tidal pools and seemed "oblivious to the other animals fleeing nearby"[3]—a clear testament to play's ability to enthrall and dramatic evidence of its risks. Play has a great many liabilities. Sharpe and other scientists believed that play would not be present in animal behavior unless its liabilities were compensated for and outweighed by adaptive advantages. Yet those advantages were, as yet, undiscovered.

The Passions of Animals

Edward Thompson's 1851 work, *The Passions of Animals*, enumerated a wide range of species that he believed engage in play. Some we might expect (such as dogs, deer, and orangutans), and others we might not (such as whales and geese). Thompson did not offer a hypothesis of play's advantage or advantages, and he made no overarching claim about play, except to say that it was a total surrender to emotion. "The animal in its sportive moments," he wrote, "abandons itself to a feeling in which its whole being seems to be concentrated in the performance of some one of its passions, whether of joy or mischief, defiance or fear."[4] German playwright and poet Friedrich Schiller opined that an animal plays from a "sheer plenitude of vitality," that play was a means for an animal to release surplus energy.[5] In 1872 English psychologist Herbert Spencer developed the idea, but it gained little traction in the natural sciences, and nothing much was written on animal play for the next several decades.

The Practice Hypothesis and the Social Bonding Hypothesis

In 1896 German philosopher and psychologist Karl Groos produced the first book on animal play, laying claim to the field with a title that was as straightforward as it was definitive: *Die Spiele der Tiere* (*The Play of Animals*). He began by cleaning house, dismissing the "surplus energy" idea on three grounds. First, a body has easier ways to release excess energy. Moreover, play is more than mere release; it has rules, form, shape. Finally, if play were no more than a means to spend surplus energy, animals would stop playing when that energy was spent—when they were tired. He noted quite pointedly that dogs do *not* stop playing when they are tired.[*]

By the time Groos was writing, the idea that play provided an animal with training had long been a piece of folk wisdom.

---

[*] "Observe the play of young dogs when two of them have raced about the garden until longing for the fray seizes him again. He approaches the other, sniffs lazily about him, and, though he is evidently only half inclined to obey the powerful impulse, attempts to seize his leg. The one provoked yawns and, in a slow, tired kind of way puts himself on the defensive; but gradually instinct conquers fatigue in him, too, and in a few minutes both are tearing madly about in furious rivalry until the want of breath puts an end to the game. And so it goes on with endless repetition, until we get the impression that the dog waits only long enough to collect the needed strength, not till superfluous vigour urges him to activity." Groos, *Play of Animals*, 19.

21

"Nothing is more common," Darwin wrote, "than for animals to take pleasure in practising whatever instinct they follow at other times for some real good."[6] But the idea had been allowed to languish. Groos turned it into a real hypothesis, proposing that animals are born with certain instincts, and that play develops them into skills necessary to their survival and reproduction— finding food, fighting rivals, escaping predators, and courting and mating. "The 'experimenting' of little children and young animals, their movement, hunting, and fighting games, which are the most important elementary forms of play, are not imitative repetitions, but rather preparatory efforts. They come before any serious activity, and evidently aim at preparing the young creature for it and making him familiar with it."[7] Groos drew upon careful observations to find common themes, anticipated and addressed possible objections, and even provided a little something extra. Natural philosophers of the nineteenth century had long wondered why the juvenile period of mammals is far longer than that of animals from other classes. Groos had an answer: a long juvenile period allows an animal ample time to play, and an animal must play if it is to develop properly. He had turned the conventional idea of play's origins inside out: "The animal does not play because he is young. He has a period of youth because he must play."[8] Groos's idea came to be called the *training hypothesis*, or more commonly the *practice hypothesis*.[9]

By the mid-twentieth century a second explanation had emerged. Called the *social bonding hypothesis*, it posited that as animals nip, tussle, and give chase, they are learning—in the phrase memorialized by kindergarten report cards—to "play well with others." Applicable at least to social animals, the hypothesis posited that many animals are born rather aggressive and asocial.

If an animal is to be accepted as a member of a group and as a mate, its aggressions must be tamed, and they are best tamed through play. Evidence for the hypothesis might be seen, for instance, in the hunting tactics of wolves. Wolves prey on animals larger than them: elk and moose. If an individual wolf expects dinner, he must hunt with a pack. But a wolf is accepted into a pack only when it has bonded with others, and it does so through play. Thus play is necessary to the survival of an individual wolf. And since a pack is composed of individual wolves, it's also necessary to the survival of the pack as a whole.

There Is No Universally Accepted Theory of Play

As these two ideas gained support, scientists studied the play of many animals—among them coyotes, grasshopper mice, squirrel monkeys, and baboons. But the research was scattershot. One study discussed sex differences in play, another contrasted play fighting and serious fighting, and another compared play to exploration. Most focused on a single species. Little progress was made toward a larger understanding of animal play. One ethologist found the situation deeply troubling.

Robert Fagen was a researcher at the Juneau Center, College of Fisheries and Ocean Science at the University of Alaska, Fairbanks. By the late 1970s he had spent fifteen years in the wilds of Alaska and western Canada observing and documenting how bears play. He was familiar enough with others' work in animal play to know the field was seeing some advances. But he also knew that before real progress could be made, researchers needed a comprehensive review of what had already been accomplished, a single work that could be used as a definitive reference. Only then could they avoid duplicating others' efforts and begin comparing results across species to build one conclusion upon another. So Fagen began to catalog animals known to play, survey the existing research, outline its questions in light of evolutionary theory, and describe the challenges researchers still faced. The result was his seminal 1981 work, *Animal Play Behavior*.

A remarkable accomplishment, the book was provocative because it made clear how much was left unanswered. With a detectable hint of lament, Fagen noted that there was no exhaustive inventory of animals that played, no agreement on what behavior should count as play, and no consensus definition of play. Perhaps most troubling was the question of play's adaptive advantage, to which there was no universally accepted theory—in truth, no real theory at all.[10]

We'll take a moment to define the word *theory*. When people say "It's just a theory" or "It's only a theory," they misunderstand the word. To scientists a theory is not trivial or facile, and they do not put one forth lightly. A scientific theory is an explanation of a natural phenomenon or set of phenomena. It is not fixed and unchanging. As more information about those phenomena is gathered, a theory that explains them is modified and improved. A *hypothesis*—another word that requires definition—is a somewhat more modest proposition, little more than a conjecture based on limited evidence. Still, a hypothesis is a beginning. Supplied with a set of hypotheses that are well-defined and testable, a scientist may begin to build a theory.

When Fagen published *Animal Play Behavior*, no single theory explained all animal play, but many hypotheses explained particular instances.

A Superfluity of Ideas

One hypothesis was put forth in the mid-twentieth century by Alex Brownlee, a Scottish researcher in diseases of farm animals. Brownlee was known to walk on a footpath from his home for twelve miles through Scotland's Pentland Hills to return a book, and along the way he'd take notes on a sheep behaving oddly or a rowan tree growing in an unusual place. In 1954 he published a paper in which he proposed that an animal playing as a juvenile exercised certain muscles, thus ensuring their proper development.[11] Brownlee never identified which muscles, and his hypothesis was never properly tested. But in the decades that followed other researchers—now grounded by twentieth-century understandings of physiology—put forth similar ideas: that juvenile play enhanced an animal's connective tissue, nervous system, or cardiovascular system. All these were specialized versions of the practice hypothesis. In 1981 ethologist Marc Bekoff, a professor of ecology and evolutionary biology at the University of Colorado, Boulder, and John Byers, a zoologist at the University of Idaho, categorized these ideas—alongside Brownlee's—as hypotheses of *motor training*.

Many thought such ideas misguided for the simple reason that play in most animals is too infrequent and short-lived to have much effect on muscles and nerves. But muscles and nerves aren't everything. Byers thought that bouts of juvenile play, while brief, might be just enough to effect changes on a more responsive and supple part of an animal's anatomy: its brain. One winter afternoon in 1993 he was deep in the stacks of his university's library, perusing for nothing in particular.[12] When he began to leaf through a textbook on neuroscience, he was stopped by a line graph. Its hill-shaped curve looked oddly familiar. He had seen it representing the frequency and duration of play over the lifetime of mice, rising to a gentle peak for a juvenile and gradually falling off into adulthood. The curve in this book stretched across the same interval, but depicted what neuroscientists call *synaptic pruning*. The cerebellums of many animals at birth have an excess of synapses, the structures through which neurons transmit electrical and chemical signals to one another. As the animal matures, in one of its "waste not, want not" moods, nature cuts away those that are little used or unused. Byers thought the overlap between the curve depicting frequency and duration of play and the curve depicting synaptic pruning might mean that play and synaptic pruning were related.

In 1995 Byers and his then graduate student Curt Walker reasoned that both changes might be brought about by a product of play: exercise. They reviewed literature on the effects of exercise in three mammals: the house mouse, the Norway rat, and the house cat. Most of those effects could occur at any stage of an animal's development, but were short-lived. Two effects—the generation of neural synapses and changes in muscle fiber—could occur only during a particular stage and were long-lasting.[13] For a period in an animal's development its nervous system and muscle tissue were permanently changed by play. Byers and Walker called their idea the *sensitive period hypothesis*. Other studies adapted the hypothesis to explain play in other animals. In 2000 Lynn Fairbanks, director of the Center for Primate Neuroethology at the University of California at Los Angeles, showed that for monkeys, different types of play peak at different stages in their development and suggested that each peak might be associated with the maturing of different parts of the brain.[14] But "might be associated with" is a world away from a demonstrable cause and effect. Even if one could be found, its scope would be limited, explaining play only in juveniles.

If play's singular function is to assist the development of the brain, nervous system, and musculature of juveniles, then we might expect those animals to stop playing when the development is complete. Indeed, many do, but by no means all—a fact that had not been lost on Groos. "A creature that once knows the pleasure of play," he wrote, "will derive satisfaction from it even when youth is gone. . . . I have a dog twelve years old that still shows a disposition to play now and then."[15] Half of all species of primates—including chimpanzees, rhesus monkeys, and humans—continue to play as adults.

The Curious Case of Cats

One might suppose that the play of adult animals *does* endow an adaptive advantage, not to them but to their offspring—that when the adults play with those offspring, they are simply being good parents. This does seem to be the case with many. For instance, adult foxes play with cubs, even ones that are not their own.[16] Yet what sometimes *looks* like adult animals teaching offspring isn't really. Consider the curious case of cats. Children who ask why adult cats and kittens play with mice before killing them are often told that the adult cats are training the kittens to hunt. This explanation is something of a white lie. In the late 1970s Tim Caro, then a PhD candidate in psychology at the University of St. Andrews, studied two groups of cats, one that had been allowed object play as kittens and one that had been denied it. To his surprise, he found no difference in their predatory skills.[17] Sarah L. Hall of the Anthrozoology Institute at the University of Southampton summarized the conclusions of similar studies, writing, "It is unclear why cats play with live prey."[18] Of scientific mysteries, this may not be ranked with the astrophysicist's "Why is 90 percent of the universe's mass invisible?," but it's surprising enough to bear repeating. Humans have probably been watching cats toy with live prey for as long as there have been humans, cats, and live prey, but we still don't know why they do it.* Thanks to Professor Caro and others, we do at least know that the reason isn't training. Since many animals play as adults with no offspring present, their play also seems to have little to do with training and parenting. An inventory of such animals would include spotted hyenas, female Steller's sea lions, male howler monkeys, lions, timber wolves, and Scottish mountain hares.[19]

By the late twentieth century it became clear that the practice hypothesis and its derivations could not explain the play of many adult animals. By the 1990s, some ethologists began to ask whether in many cases play's adaptive function wasn't long term at all. Perhaps it was short term. Perhaps it was immediate.[20]

In the early nineties Irish-born professor of psychology Nigel Barber noted that since most newborn mammals have one or more parents providing for them, they don't need to hunt or

---

* For Darwin, the cats' evidently sadistic behavior was evidence against intelligent design: "I cannot persuade myself that a beneficent & omnipotent God would have designedly created the Ichneumonidæ with the express intention of their feeding within the living bodies of caterpillars, or that a cat should play with mice." Darwin Correspondence Project, letter no. 2814, accessed January 18, 2018, http://www.darwinproject.ac.uk/DCP-LETT-2814.

forage. Consequently, they *do* have energy to spare. Barber found that play, and vigorous play in particular, increases metabolic rates and enables *thermogenesis*, the burning of brown adipose tissue.[21] Thermogenesis warms a mammal and simultaneously enhances its immune system. Both are immediate benefits. Thus the surplus energy hypothesis of Herbert Spencer, an idea roundly dismissed by Groos and largely forgotten for nearly a century, was like a long-lost ball found, reinflated, and returned to the game.

Barber's hypothesis was limited. It explained some newborn and juvenile play well enough, but it explained adult play only in mammals that burn brown adipose tissue—that is, mammals that hibernate. It did not explain adult play in many other mammals—among them adult kangaroos, rats, primates, and seals. And it did not explain *any* play in birds or other nonmammals. So while the hypothesis was persuasive, it was a long way from a comprehensive theory of animal play.

In 1998 ethologist Katerina V. Thompson drew on a number of studies to make a case that animals play not so much to improve their abilities as to assess them, like athletes acting as their own coaches.[22] This resembled Groos's practice hypothesis, slightly revised, updated, and supplied with scaffolding. If an animal fails to complete a certain action, Thompson wrote, it may abandon the effort or try again. If upon trying again it succeeds, the animal may then attempt something more difficult.

An Inquiry Unraveling (or Maybe Just Unraveled)

By this time the many answers to the question of the adaptive advantage of play were disputed, and none were comprehensive.

If you think that the inquiry was becoming a bit frayed, you'd have been in good company. In 1991 Barber made his displeasure clear, writing, "Play research suffers from the inheritance of empirically unsupported assumptions that are frequently accorded the status of fact."[23] They are accorded that status only because they "satisfy a need for explanation, and have been repeated by authorities."[24] Thompson was equally blunt: "The play literature abounds with hypotheses and speculation, but quantitative support for most functional hypotheses is in frustratingly short supply."[25]

As previously noted, many ethologists assumed animal play had one or more adaptive advantages precisely because it had so many obvious *dis*advantages. Through the second half of the twentieth century that assumption had hardened into orthodoxy, and yet it was unproven. Never mind *how* play increased an animal's chances for survival or reproduction. No one had shown that it increased those chances in the first place.

Some thought that play was possibly just a feature, like the human appendix or the panda's thumb, that had no adaptive advantage. As such, it would not be unusual. Natural selection often brings about adaptations slowly, in a piecemeal, makeshift fashion. When habitats change, even adaptations that seemed "most fit" may suddenly be rendered useless. It should not surprise us, Darwin noted, "that there should be geese and frigate birds with webbed feet, either living on the dry land or most rarely alighting on the water; that there should be long-toed corn-crakes living in meadows instead of swamps; that there should be woodpeckers where not a tree grows, that there should be diving thrushes, and petrels with the habits of auks."[26] Play might be no more advantageous to a squirrel monkey than webbed feet were to a frigate bird living on dry land. Differences between populations that play and those that don't might also be due to nothing more than sampling or experimental error. But such ideas rarely figured in ethologists' thinking about play. Many believed strongly that animals' recreational activities helped them to survive and reproduce. Yet there was no evidence.

A Step Forward: Play and Survival of Free-Ranging Brown Bears

In the first decade of the twenty-first century, to the collective relief of many, several ethologists found some. In 2004 behavioral ecologist Scott Nunes and colleagues discovered that female Belding's ground squirrels that engaged in social play produced more offspring than those that did not. And in 2008 wildlife biologist Elissa Cameron and her team found that among feral horses, foals that spent more time playing were more likely to survive their first year of life.

Among other studies, perhaps the most ambitious was undertaken by Robert and Johanna Fagen.[27] Their subjects were free-ranging brown bears living on Admiralty Island, in Alaska's Alexander Archipelago. One locale particularly attractive to the bears is Pack Creek, a salmon stream that runs through old-growth spruce and hemlock forest and then divides into channels to form a long estuary. In July 1985 the Fagens began observations here that would continue through the next nine summers.

Identifying the bears' play was not difficult. When brown bears play with objects, they take on a characteristic posture with a raised shoulder, foreleg, and paw. Their social play is likewise quite recognizable: chasing and wrestling, easily differentiated from real fighting because it's undertaken in silence.

The Fagens knew that many factors might aid any cub's survival: its health, the degree of protection and nurturing supplied by its mother, and the quality of its nutrition. The Fagens took pains to distinguish these from the effects of play. And they were careful to ensure that no single mother in the study had an outsize influence on the results.

They studied eleven families and nineteen cubs, observing each cub for at least thirteen hours over three months every summer for ten years. One cub died during the first summer. Others did not return in the second summer, having presumably perished from disease, hypothermia, or malnutrition. Only eleven of the nineteen survived to the beginning of their third or fourth summer, a rate of mortality that, while shocking to us, is fairly typical for brown bears. Play, however, seemed to reduce that rate. The Fagens found that most of the bears who survived year to year were those that had played more as cubs.

Correlation does not equal causation; the Fagens allowed that the surviving bears might have survived for reasons other than their play. Still, it was careful work and something of a model. Their findings, like those of Nunes and Cameron, showing a clear correlation between play and survival, were important. At long last there was evidence, if not definitive proof, that play provided real adaptive advantages.

Now it was a matter of discovering what those advantages were.

The Kalahari Meerkat Project

When Lynda Sharpe was a child, she had listened to Kipling's *Just So Stories*—"How the Leopard Got His Spots," "How the Camel

Got His Hump," and the like. Their whimsy delighted many but left her, as they did many a budding evolutionary biologist, unsatisfied. Exactly how a leopard got its spots was a good question. Even as a child, she knew that it must have a real answer, and Kipling was of no help finding it. She kept asking questions, and years later as an undergraduate at Monash University in Australia, she focused those questions on play—particularly that of captive African wild dogs and carnivorous marsupials. By 1995, she was a doctoral candidate at Stellenbosch University in South Africa with a research emphasis in zoology. Like Fagen and Barber and Thompson, Sharpe was dissatisfied with the lack of progress in the study of animal play. She decided that her dissertation would include long-overdue tests of both reigning hypotheses—that is, the practice hypothesis and the social bonding hypothesis.

Both hypotheses had long been part of folk wisdom and seemed commonsensical. But even by the last decade of the twentieth century, neither had been subject to rigorous tests. The reason in part was owed to a dearth of places in which such tests might be performed. Studies of animal behavior are undertaken in one of two settings, and neither is ideal. Researchers working in laboratories can observe their subjects easily and, by manipulating conditions in their environment, conduct controlled experiments. But laboratories are artificial environments and likely to cause animals to adapt or otherwise adjust their behavior, thus rendering the conclusions of any observer open to question. Researchers studying animals in the wild, on the other hand, may find their subjects difficult to observe for long intervals, and the more timid difficult to observe altogether. A researcher's preferred sort of animal was one that lived in its natural habitat and could also easily be observed.

Sharpe knew of an animal that met both conditions: a species of mongoose called meerkats. Better yet, she knew where to find a great many. Several groups, each with twenty to fifty individuals, inhabited South Africa's Kuruman River Reserve, in the southern Kalahari. They were already subjects of the Kalahari Meerkat Project, a long-term research effort focused on animals' cooperative behaviors.[28] Sharpe joined the project in 1996. But how should she begin her own work? There was no template for tests of the practice hypothesis or the social bonding hypothesis, and no testing methodology that might be applied across species.[29] Designing such tests would take imagination, and their implementation would require patience.

The Practice Hypothesis Tested

The practice hypothesis posits that through play an animal develops skills necessary to survive and reproduce, to achieve "evolutionary success." Sharpe knew that one easily measured index of such success is the number of offspring. She also knew that the meerkats with the most offspring in any group were the dominant pair, and that they achieved their dominance by winning fights against other meerkats.[*] Thus it followed that winning fights led to evolutionary success. But how did the dominant pair learn to win fights? Like all meerkats, they had play fought as pups. Was their play fighting good practice for winning actual fights? Was it even good practice for winning play fights? Sharpe took on

---

[*] The dominants' rule can be merciless and brutal. The dominant pair may exile members and kill the offspring of others, thus ruthlessly ensuring that their own offspring will meet with no competition.

the second question first. She would ascertain whether meerkat pups who play fought most also won the most play fights.

To make her work easier, Sharpe gave the meerkats visible, distinguishing marks. Since they had already grown used to researchers' presence, this was not difficult. For several mornings she arrived at a meerkat burrow before dawn and waited. Soon a group would emerge from the burrow, and as they yawned, stretched, and warmed themselves in the morning sun, she would crawl among them, catch one, and with a felt-tip pen draw a ring on the only body part she could reliably make out in the tumult of a play fight—the tail. She used a different color for each individual. Then she'd sit and wait. A few would frolic alone, others would pull at a leaf or a twig, and still others would tussle with one another—that is, play fight. A bout began when two, standing on hind legs, used their upper bodies to push against each other, clasping with claws and nipping each other's head or neck. Sooner or later one would push the other onto its back and stand over it with hind legs on the ground, pinning it with forelimbs. Sharpe judged the moment of pinning to be the end of the play bout, and the meerkat on top to be its winner. Sometimes the pinned meerkat would wriggle free and escape, and the other would give chase. Sometimes the meerkat on top would simply release the one pinned.

In time she counted an impressive 27,100 play bouts between thirty-seven meerkat pups (most having skirmished more than once) from six groups. Thanks to those felt-tip-pen rings, she was able to track many through their first year of life. Within a few years she had reams of data—tallies of the play frequency of individual meerkats over time and of groups by age, win-loss records of individuals and groups, and enough charts, bar graphs, and bell curves to satisfy the most demanding statistician. The upshot? Perhaps surprisingly, perhaps disappointingly, the meerkat pups who did the most play fighting were *not* the ones who won the most play fights.

And so to the other question. Did the meerkats who won the most play fights as pups win the most real fights as adults? She began to count actual fights. Some she witnessed firsthand; others she surmised from bite wounds. When she compared win-loss records of each adult with the play-fighting win-loss records of their youth, she found that meerkats that play fought more in youth were no more likely to win real fights as adults, so no more likely to become dominant, and so no more likely to produce offspring.[30]

The practice hypothesis, at least as applied to play fighting in meerkats, seemed wide of the mark.

Sharpe's project had a second part. During the six and a half years when she observed her "little darlings" as a test case for the practice hypothesis, she also observed them as a test case for the social bonding hypothesis. We'll recall that its advocates posited that bonds between an individual animal and the animals with which it lives are forged in play. Meerkats made for an especially expedient test of the idea. To understand why, we need to know something of meerkat social bonding and of the environment in which it evolved.

Survival in the Kalahari

The Kalahari is not a true desert, but rather a semiarid savanna. Meerkats there face almost continual danger. Martial eagles—Africa's largest bird of prey—circle overhead, jackals prowl among the grasses and sand dunes, and the ground is home to several species of venomous snakes, the deadliest being Cape cobras and puff adders. Each meerkat group ranges over a territory of several square miles, and neighboring groups war. All these dangers compel any meerkat group to protect its own. When the group's members are outside their burrow, one acts as a sentinel. If the sentinel sees a predator, it barks or whistles, and the others stop whatever they're doing and scurry into their burrow or—if they are too far from it—into the shallow trenches spread across their territory that Sharpe and others call bolt-holes, dug in anticipation of the need to hide.

This all-for-one, one-for-all approach to life is especially interesting to students of evolutionary biology because on geological scales it's a recent development. The ancestors of present-day meerkats who lived during the Pliocene, the geological epoch lasting from 5.2 million to 2.64 million years ago, foraged alone. Much of the Kalahari then was tropical forest, and its dense undergrowth provided an animal cover. But as the climate dried out and the forest gave way to open savanna, a lone animal became easy prey. If those ancestral, Pliocene meerkats were to survive in the new, harsher environment, they had to overcome at least some of their unsociable tendencies and learn to cooperate. Clearly, that's exactly what they did. But how? How did animals that were solitary by nature come not only to watch out for one another, but to develop elaborate protocols for doing so: standing guard, protecting one another's young, sharing food, and sometimes tidying up the place? As we've seen, meerkat play is a decidedly social endeavor. Might it be the means by which natural selection turned a solitary animal into a cooperative one?

The Social Bonding Hypothesis Tested

The bonds of any meerkat group were evident in meerkat civic duties: acting as sentinels, clearing sand from bolt-holes and sleeping burrows, sharing food with young on foraging expeditions, and minding pups. It followed, Sharpe thought, that she would have support for the social bonding hypothesis if she could show that the meerkats that played the most as pups were those that as adults undertook these duties most readily and most often. Yet she found none. The blond meerkat Mimi and a meerkat named Goblin played a great deal as pups, but upon reaching adulthood neither took on more responsibilities than those who played less. The male cub Bandit played little, yet he became the group's most dedicated pup minder. Upon examining the results of this study in its entirety, Sharpe could see no connection between play and good citizenry. She made several other tests of the social bonding hypothesis, some involving group size, others grooming, and still others long-term alliances between individuals. None of the results showed that play helped meerkats bond.[31]

Here then was an eminently playful animal for whom practice and social bonding—the two prevailing hypotheses to explain animal play—simply didn't apply. For meerkats, both hypotheses turned out to be their own kind of *Just So* stories. Disappointing results, but still significant, especially for the social bonding hypothesis. Since bonding is crucial to meerkat survival, one might expect that they'd use any and all available means to achieve it. That meerkats had play in their behavioral repertoire and yet did *not* use it to develop social bonds suggests—and we must say here *only* suggests—that other animals for whom social bonding is not nearly so critical probably didn't either. "Despite more than three decades of research, and the postulation of more than 30 hypotheses of function," Sharpe wrote, "the adaptive significance of play remains unknown."[32]

Natural Selection, Like Play, Exacts Great Costs

Play is famously profligate. In fact, its immoderate use of an animal's time and energy is what prompted ethologists to wonder at its adaptive advantage, and Sharpe to undertake so many studies of meerkats. That profligacy has a cognate in natural selection. As Darwin noted, "Many more individuals of each species are born than can possibly survive."[33] That excess is necessary to a long-term benefit. The greater the number of individuals born, the greater the number born with adaptive advantages, and the greater the number that, by virtue of those advantages, will survive long enough to reproduce, pass on those advantages to their offspring, and thus make an evolutionary line "more fit."

Some years before Sharpe began observing meerkat play, three ethologists—two Americans and a Czech—suggested that the profligacy of play may also be necessary to a long-term benefit. This benefit is not to an evolutionary line, but to an individual animal. Their idea was inspired by a curious move in the play of a small mammal that lived not in the Kalahari, but in a place quite unlike it: the hilly, mist-shrouded, and often rainy terrain outside a city some eight thousand miles to the north.

CHAPTER 3

Tumbling Piglets and Somersaulting Monkeys: Training for the Unexpected

In the heart of Edinburgh's Old Town just outside Greyfriars Kirkyard is a granite pedestal atop which sits a life-size bronze replica of a Skye terrier. Most days you'll see people gathered around it taking pictures, and sooner or later a parent will lift a child so she might touch its nose for luck. The object of their veneration is Greyfriars Bobby, who is believed to have kept vigil over the grave of his master from 1858 to 1872. The statue is a tribute to that fidelity, and a fitting civic symbol for a place with a centuries-long history of humans' concern for, and love of, our fellow creatures.

If you turn from the statue and walk southward along the curving, shop-lined streets of Lothian and Potterrow, you'll soon come to the grounds of the University of Edinburgh, where, some years before signing on as unofficial naturalist aboard the HMS *Beagle*, a young Charles Darwin studied medicine. It's the same university that employed Darwin's protégé George Romanes as a professor, the university from whose veterinary college Alex Brownlee earned a degree, and the university that supports the world-renowned Roslin Institute, where in 1996 Ian Wilmut, Keith Campbell, and their colleagues cloned a sheep they named Dolly.

In the 1980s you could also venture a bit outside the city to an open hillside, 230 acres of grassy slopes, streams, and a small wood. This was the Edinburgh Pig Park, a place where domesticated animals could roam freely and still be easily observed and studied. It was the brainchild of an instructor in the university's Department of Agriculture named David Wood-Gush. As a young man, Wood-Gush had sustained injuries from a motorcycle accident severe enough to necessitate the amputation of part of his left arm. For much of his life he endured phantom pain. This discomfort, he would say, instilled in him an empathy for domestic animals, and inspired him to seek methods that might lessen their suffering.

In 1988 the Pig Park was home to five adult sows, an adult boar, and a great many piglets. This was when Wood-Gush and his colleague Ruth Newberry undertook a study of the play of pigs, a subject that they believed was much in need of attention. Play had proven to be a fair indicator of the health of pigs (healthy pigs play more), but no one had described the extent of their play in all its varieties, much less quantified it. Such knowledge would help to devise conditions under which pigs might be raised humanely.

In their research, Newberry noticed some curious porcine behavior. When the piglets played, they often ran about. This, she thought, was the sort of play easily explained by evolutionary biology. Running had an obvious adaptive advantage: it would be good practice for escaping a predator. But at no particular time and for no apparent reason, one piglet would suddenly stop running and perform a flop-over. This certainly looked like play, but for what? Coming to a full stop only to tumble arse-over-teakettle would seem to be good practice for nothing—except perhaps for making yourself a predator's dinner.

Some years later, Newberry took a position at Washington State University, where she collaborated with Marc Bekoff, then at the University of Colorado, Boulder, and Marek Spinka, also in Boulder, visiting from the Research Institute of Animal Production in the Czech Republic. Each had the humane treatment of animals domestic and wild as their primary research area, and each had also accumulated years of observing animals in their natural environments. Together, they called upon that knowledge to devise a new hypothesis of play's adaptive advantage and, not incidentally, an explanation of that piglet's flop-over.

Getting Real

Free-ranging animals don't run on a smooth, well-groomed track. They run over uneven and perhaps slippery ground, a terrain likely to be strewn with obstacles. Some of those obstacles, such as roots and stumps, are inanimate. Others, such as animals running alongside, decidedly not. All of which means that, while running, free-ranging animals are likely to stumble, slip, fall, or collide with something. Spinka, Newberry, and Bekoff knew that the piglet flop-over was not good practice for escaping a predator in an idealized environment. It might, though, be good practice for recovering from a fall in a real one. Natural selection might have developed a means for animals to learn to recover balance by evolving in them a desire to put themselves in situations where they will be thrown off-balance. "We hypothesize," they wrote, "that a major ancestral function of play is to rehearse behavioral sequences in which animals lose full control of their locomotion, position, or sensory/spatial input and need to repair those faculties quickly."[1] They called their idea "training for the unexpected."*

---

* While Spinka and company held that training for the unexpected was likely the original adaptive advantage of play, they allowed that others have evolved since. "While the core of our hypothesis is that play in mammals has one original function, to train for unexpected situations, clearly play behavior in its variety can fulfill various other functions in individual species and animals that differ in age and sex." Spinka, Newberry, and Bekoff, "Mammalian Play."

Training for the Unexpected

The piglet, like most active animals, has a set of standard body positions and "species-typical motor patterns," that is, positions and actions to which it is accustomed and uses often. A piglet knows how to couple one action with another, and that to still another: scampering, turning, stopping, etc. But a piglet catching a hind foot on an exposed root, stumbling, and falling is unlikely to find itself in a standard body position, and it has no species-typical motor pattern to get back to that position. It must improvise, inventing *new* moves—a sideways twist perhaps, along with a kick from a hind leg. And so it does, adding to its behavioral repertoire. Training for the unexpected, the researchers suggested, was enabled by a dialogue between the parts of the central nervous system concerned with muscular action.

They suspected that it also involved the parts of the brain that control emotions. A piglet suddenly finding itself fallen on the ground is likely to feel frightened, and fear may provoke panic and a consequent loss of control over motor functions that may make the piglet—already vulnerable to predators—even more vulnerable. If a piglet, or any other animal, is to survive the unexpected and improvise new moves, it must maintain control over those functions. To do that, it must suppress its fears and remain calm.

Here things get more complicated. Consider the moments before the flop-over as the piglet experiences them. We may presume that the piglet does not say to itself, "Although I am scampering in what would be an efficient manner in an idealized environment, the environment in which I am actually scampering is marked with features that inhibit or interrupt movement. So that I might prepare for moments when my movement will be inhibited or interrupted and from which I might learn to recover, I should simulate such a moment right now, with a flop-over." Rather, the piglet performs the flop-over because it feels good, and because it's fun. The craving for fun is not the ultimate reason for the flop-over or the reason natural selection chose it. It is, however, the immediate reason the piglet chose it.

A Close Look at Fun

Spinka, Newberry, and Bekoff maintain that an animal begins to play because it seeks excitement. But it will seek excitement only if it is free of stress and anxiety—if it is relaxed. If the animal is to stay relaxed, the play cannot be dangerous; yet if the animal is to continue to play, it must stay excited. Play, then, is exciting, but not so exciting as to be dangerous. It is *thrilling*.

Further, they maintain that any moment of play may be divided into a sequence of shorter moments, some more gratifying than others. For the piglet, the flop-over necessitates a loss of control and a measure of risk and uncertainty likely to make it feel anxious. The piglet undertakes the flop-over not for its own sake, but in anticipation of the moment immediately *after* the flop-over when it recovers and regains control. That part of the experience is rewarding and keeps the piglet playing.

The catch here is that the moment of recovery is far too brief for an animal that wishes to keep playing. The animal cannot sustain that moment, but it can do the next best thing: re-create it. If an animal is to recover control, it has to lose it first. So an animal at play deliberately surrenders control and recovers it, over and over, producing the mildly addictive "more, again, now" sensation. We all know and probably enjoy the feeling. It's why gamblers gamble, why roller coasters roll, and why novels and movies have rising and falling action. It's why Sam plays it again.

Self-Handicapping

Spinka and his colleagues called that surrender of control *self-handicapping*. They thought it an essential feature, perhaps the most essential feature, of play.

The piglet flop-over was self-handicapping by a single animal in one mode of play: that is, a piglet in solitary play. But training for the unexpected is versatile, the utility player of the play hypotheses. It can be extended to explain, for instance, object play. Mather had posited that an animal investigating an object is gathering information about it, while an animal playing with an object is discovering what can be done with it. In identifying self-handicapping as an essential feature of play, Spinka and his colleagues sharpened that distinction, noting that while an animal that is investigating does not self-handicap, an animal at play often does.*

One of the strengths of the training-for-the-unexpected hypothesis is that it supplies a means to recognize play. Spinka and his colleagues noted that movements in serious behavior—ones

---

* Moreover, Spinka and his colleagues noted, self-handicapping highlighted the adaptive advantage of play and distinguished it—also quite clearly—from the adaptive advantage of exploration. Exploration alerts an animal to threats and helps it *avoid* trouble. Play offers practice for uncertain situations and helps an animal get *out of* trouble.

used to escape predators, chase prey, and so on—are efficient, spending as little energy as possible to achieve their purpose. If the movements of play were directed toward the same ends, they would be inept and ineffectual, in their words, "too disproportional, too distorted, too fast, too uncontrolled, or too quickly repeated."[2]

A website devoted to *Canis familiaris* calls the American leopard hound a breed that is "eager, loyal and fast." I knew one who was certainly eager and loyal, but fast only when he wanted to be. He would often lie on his side and in an ungainly manner slowly push a rag doll across the floor, taking several minutes to move it a few feet. Spinka and colleagues would not be surprised by this behavior. They'd say that the dog wasn't interested in moving the doll. Rather, he was interested in restricting his own range of movements so that he might enjoy a new experience. Seen in the framework of evolutionary biology, the dog was learning to manipulate objects in new ways, from unusual angles and disadvantaged positions. Dogs are not alone among self-handicapping object players. Many species of primates—including lemurs, marmosets, and monkeys—manipulate objects while hanging upside down from their tails or feet.

Self-handicapping can be further extended to explain social play. Animals who play fight—such as wolves, dogs, kangaroos, and rats—often self-handicap. One of the play fighters will deliberately fall backward, putting itself in a decidedly disadvantaged position. It's a fairly curious move. Ethologist Maxeen Biben proposed an explanation for it. In a play fight whose participants are not well matched, the stronger partner uses self-handicapping to make the play less threatening for the younger or weaker play partner. It does so not to win the play fight, but to ensure that it continues.[3] Spinka and company affirmed Biben's hypothesis and framed it within their own, positing that the stronger animal in a play fight self-handicapped for the same immediate reason an animal self-handicapped in solitary play or object play—it was pleasurable. The ultimate reason was that, like self-handicapping in any form of play, it offered training for the unexpected.

All sounds reasonable enough. There is, though, a complication. In many play fights, the younger and weaker partner *also* self-handicaps, a behavior that most hypotheses of animal play had little or nothing to say about. Spinka and company posited a subtle communication between the players. When the weaker self-handicaps, it is signaling that if the stronger wishes to extend the play, it must restrain itself. The stronger takes the hint, falls backward, the weaker pounces, and the play goes on.

And what of those meerkats? We'll recall that young meerkats who engaged in more play fights than their peers did *not* win more often, and that meerkats who play fought more in youth were no more likely to win real fights as adults. Lynda Sharpe had referenced the article by Spinka and his colleagues and so knew of the self-handicapping hypothesis, but proceeded under the supposition—the same supposition of Katerina Thompson—that when animals play fought, they play fought to win. Spinka and colleagues would suggest that those meerkats were *not* playing to win.

The training-for-the-unexpected hypothesis posits that play, whether that play is solitary, object, or social, provides many benefits. When confronted with the unexpected, an animal that plays will experience a less pronounced physiological stress response and recover more quickly from a collision or fall. It will be less likely to panic when escaping from predators and will resolve conflicts more easily and thus suffer fewer injuries from those conflicts. Lastly, upon finding itself in a new environment, it will move from exploration (a relatively anxious emotional state) to play (a relaxed one) more quickly.

Of the thirty-odd hypotheses of animal play, some are little more than notions. They make no predictions about what features in an animal's behavior scientists should expect to find. The more developed hypotheses—those that do make predictions—address a range of features: the nature of movements during play, how an animal's environment affects its play, play's role in an animal's developmental stages, and play's effect on an animal's brain and nervous system. Most of these hypotheses make only a few predictions, many of which are difficult to test. A virtue of the training-for-the-unexpected hypothesis is that it makes a great many predictions, and most are testable.

Predictions of the Training-for-the-Unexpected Hypothesis

Animals intent on achieving a goal are likely to avoid mediums that restrict their mobility. The training-for-the-unexpected hypothesis predicts that animals that are playing or intending to play, and thus are in pursuit of the unexpected, may seek out such mediums. This seems to be the case. Siberian ibex kids, for instance, have been seen to prefer playing on sloped ground, and bighorn sheep in sand.[4] A more familiar example is a puppy introduced to new-fallen snow: a moment of uncertainty, a brief hesitation, then a complete surrender to the impulse, a leap—a leap of faith if there ever was one—and finally a joyful romping in and through the snow. The puppy's pleasure is self-evident but, for many hypotheses of animal play, difficult to explain. The puppy will find its movements inhibited and, if the snow is deep enough, its vision compromised. How can that be fun? Spinka, Newberry, and Bekoff would say that the reason is clear. Snow handicaps the puppy or, rather, since the puppy chooses to make the leap, snow offers an opportunity to self-handicap, which the puppy seizes because self-handicapping is fun.

The hypothesis predicts that to increase opportunities for disorientation and recovery, animals at play will switch rapidly from one kind of movement to another, squeezing a lot of disorientation and recovery (and therefore a lot of training for the unexpected) into a brief interval. The self-handicapping of that American leopard hound was much like this. He pushed the rag doll only a short way before he jumped up, ran around the room, gazed briefly at a movement outside a window, returned to the doll, and pushed it a bit farther.

For a playing animal, what's better than playing in a single novel medium is moving quickly in and out of two or more mediums. Groos noted the behavior in harbor seals near San Francisco: "They may be seen plunging into the sea, either sliding down the smooth, sloping sand banks or throwing themselves from a high rock; then they carry on their play like dolphins, rapidly throwing themselves over so that the belly is uppermost, and sometimes springing entirely out of the water."[5] The seals were using their variegated surroundings to pack a lot of disorientation and recovery into a brief interval. As training for the unexpected, the behavior was efficient indeed.

The hypothesis further predicts that animals playing together don't need a novel medium to encounter the unexpected. They may introduce it themselves, as sooner or later one will make a move that the other does not anticipate, thereby offering it an opportunity to improvise. If three animals are playing, then each animal's opportunities to improvise are that much greater, and the chances that each will encounter the unexpected are greater still. The more animals play, the more unexpected each animal encounters, and the more opportunities each has to train for it. Moreover, an animal seeing others at play is assured that it has found a sweet spot—enough unexpected to be exciting, but not so much as to risk injury. For these reasons, Spinka and company posited that animals will prefer to play with more than one partner, and an animal seeing others at play will want to join in. In short, play should be contagious. This often seems to be the case: Sharpe's meerkats and Newberry's piglets had a more-the-merrier take on their play, and the same seems true of hyenas and Steller's sea lions. If the snow-smitten puppy saw another puppy already romping in the snow, it probably wouldn't hesitate at all.

Yet some animals readily engage in play, even play that is clearly dangerous, with no assurance of their safety. Two longtime animal-play researchers were surprised to come upon a group of animals not only risking injury, but actually *inviting* it.

The Riddle of the Belly-Flopping Monkeys

Sergio Pellis and Vivien Pellis are faculty at the University of Lethbridge's Department of Neuroscience in Alberta, Canada. They are also spouses and frequent coresearchers in animal play. In the late 1990s the Pellises, partnering with a zoo in nearby Calgary, undertook a study comparing the play-fighting styles among three species—ring-tailed lemurs, black-handed spider monkeys, and patas monkeys. The Pellises positioned video cameras in the monkeys' habitats and, over several days and weeks, monitored and recorded the monkeys' behavior. When viewing the footage, they saw something they hadn't expected.

A group of juvenile patas monkeys—a species of primate native to central Africa with reddish-brown fur, black faces, and long tails—were scrambling around on the ground near a large tree. Then three of them ran up the tree's trunk, and one after another jumped from a branch toward a grassy patch of ground. Monkeys must learn to swing from branch to branch safely, and that learning can come at some cost: autopsies of monkeys commonly reveal bone fractures. Monkeys must also learn to land safely—and that's where things in the Calgary zoo got a bit strange. A monkey learning to land without hurting itself will curve its back and prepare to strike the ground with its feet, distributing the shock upward through its legs and spine. But the patas monkeys didn't do this. Instead, after launching themselves from the branch, they spread-eagled and belly flopped onto the ground with what the Pellises described as "sickening thuds." The hard landings were not accidental. Far from it. Each monkey, after taking a moment to regain his composure, climbed the tree and did it again—at least ten more times.[6]

The Pellises had spent much of their careers observing animal play, but they had never seen anything quite like this. It was as though they'd happened upon a patas monkey subculture of masochists. If the adaptive advantage of play generally was a mystery, this particular instance brought that mystery into sharp relief: the evident *disadvantage* of the play here was extreme. Serial belly flops onto hard ground didn't merely risk pain and injury. They all but ensured it. Yet this, too, may have been training for the unexpected—the monkeys seemed to be getting used to pain, and learning to recover from it.

Notably, the belly-flopping monkeys were juvenile males. In many mammal species male juveniles are the most playful. Dangerous behavior among juvenile male *Homo sapiens* is so familiar as to be a cliché. It's why, when I was found to have followed someone else into some no doubt unwise youthful pursuits, an elder would ask the ancient rhetorical question "If Jimmy jumped off a bridge, would you?" The honest answer, the one I knew better than to give, was "Maybe."

Dogs, Chamois, Elephants, and Darwin's Children: Large-Mammal Sliding

A gentler surrender to gravity, a slide, seems the preferred recreational activity of certain animals. Dogs are known to take pleasure in cavorting on snowy slopes, sliding down and bounding up again for another go. Groos described the same behavior in chamois, the agile goat-antelope that inhabits mountainous regions from Spain to the Caucasus:

> When in summer the chamois climb up to the perpetual snow, they delight to play on it. They throw themselves in a crouching position on the upper end of a steep, snowcovered incline, work all four legs with a swimming motion to get a start, and then slide down on the surface of the snow, often traversing a distance of from a hundred to a hundred and fifty metres in this way, while the snow flies up and covers them with a fine powder. Arrived at the bottom, they spring to their feet and

slowly clamber up again the distance they have slidden down. The rest of the flock watch their sliding comrades approvingly, and one by one begin the same game. Often a chamois travels down the snow slide two or three times, or even more. Several of them frequently come roughly together at the bottom.[7]

There may be a whole behavioral category of large mammal sliding. For some large mammals, when snow is unavailable, another substrate will serve just as well. Many have reported a seemingly common behavior among pachyderms in parts of Indonesia: an impromptu session of slip-and-slide. A young male elephant was said to wait high on an embankment as two others climbed the slope. When they were halfway up, he sat on his haunches, slid, and collided—it seemed deliberately—with one of the others, who then slid with him. Upon reaching the bottom, they both headed back up the embankment, but the third elephant was now sliding down the slope. He collided with them, and all three ended at the bottom in a great muddy heap.[8]

The affinity for sliding is well-known in human children, and Darwin, a devoted father to ten, made efforts to satisfy it. Among the furnishings of their home, the Georgian manor named Down House, was a portable wooden slide that could be laid over one side of the main stairway so that the children might enjoy the mild exhilaration of a slow, controlled fall.

Play and Developmental Stages

Yet another prediction of the training-for-the-unexpected hypothesis is specific to an animal's developmental stages. Roughly speaking, a mammal experiences three such stages, being a *newborn* from birth until it weans, then a *juvenile* until it reaches sexual maturity, and then an *adult* thereafter. Among most mammals, newborns play some, juveniles play a lot, and adults play a little or not at all. The training-for-the-unexpected hypothesis predicts that if you graphed the amount of time mammals devote to play in a given stage, you'd find that for most the line for solitary play would peak first, object play next, and social play last.

If playing animals are training for the unexpected as they are developing, so say Spinka and company, we should not be surprised to find that solitary play peaks before object play. An animal needs to be able to control its movements before it can effectively manipulate things in its environment. We should also not be surprised to find that solitary play peaks during the juvenile stage, when an animal's body parts are developing quickly and at different rates. It's known as "that awkward age" in teenage or preteenage humans, and as those of us who survived it may recall, a lot then was unexpected. Finally, we should not be surprised to find that social play peaks last. By the time an animal reaches adulthood it has gained control of its body, which is no longer developing and has become familiar. It has also gained control over many things in its environment. But not all of them. It has not gained control over the behavior of others, and never will. But through social play, it can learn to expect and prepare for it.

The hypothesis further predicts that an investigation of adult play would show that most of it is social. This seems to be the case for a great many adult animals—among them bighorn sheep, dwarf mongooses, orangutans, and bonobos.[9] The training-for-the-unexpected hypothesis, unlike the practice hypothesis and some of its more recent variations, supplies a satisfactory answer to the question of why adult animals continue to play.

A Way to Categorize Play—the Ethogram

Some fields of science have a ready means to organize and classify their subjects. Chemistry, to take the obvious example, has the periodic table, a conceptual cabinet of well-ordered and tightly walled compartments, a place for every element and every element in its place. A chemist can be certain, for instance, that carbon is always to be found in the second row from the top, slotted neatly between boron and nitrogen, and that it will stay there as long as the universe has carbon. But evolutionary biologists, for whom the only constant is change, are not as fortunate. They have no classification scheme so permanent, and none is possible. Name a thing, fix it beneath your semantical specimen slide, and given enough time it is sure to wriggle free, and you'll need to name it again. Evolutionary biology does have an open-ended classification scheme, one that biologists these days call the phylogenetic tree, and that Darwin called "the Tree of Life."* Another scheme used to impose rule on the

---

* Many have used the metaphor, but Darwin's own explication can hardly be improved upon and rewards rereading: "The green and budding twigs may represent existing species; and those produced during each former year may represent the long succession of extinct species. At each period of growth all the growing twigs have tried to branch out on all sides, and to overtop and kill the surrounding twigs and branches, in the same manner as species and groups of species have tried to overmaster other species in the great battle for life. The limbs divided into great branches, and these into lesser and lesser branches, were themselves once, when the tree was small, budding twigs; and this connexion of the former and present buds by ramifying branches may well represent the classification of all extinct and living species in groups subordinate to groups." Darwin, *Origin of Species*, 113. It's a vast, expansive, and protean image for a vast, expansive,

unruly, and indispensable in the field of animal behavior, is the *ethogram*—a chart of kinds of behavior in an animal or animals.

By the first decade of the twenty-first century, ethologists using Burghardt's criteria had identified playful behavior in several species of primates—macaques, chimpanzees, lemurs, and bonobos, among others. The playful behavior in each resembled serious behavior to greater and lesser degrees. An ethogram that highlighted these degrees would allow ethologists to better understand the relation of play to serious behavior in each species, and to compare play behaviors across these species. Yet no such ethogram existed. Likewise in 2001, when Spinka and company published their paper positing that play was training for the unexpected, no ethogram had distinguished play that was self-handicapping from play that was not. An ethogram showing that little or no play was self-handicapping would mean that their hypothesis might need rethinking. But an ethogram showing that a significant part of all play was self-handicapping would be evidence that play—much of it anyway—was training for the unexpected.

The Somersaulting Langurs of Bhangarh, Rajasthan: A Test of Training for the Unexpected

In the first decade of the twenty-first century, Milada Petrů, a zoologist at Charles University in Prague, and a research team that included Spinka began a study to test the predictions of the

---

and protean subject. Since evolutionary biology informs and underlies each of the life sciences, the phylogenetic tree, or a version of it, can be called upon to represent physiology, anatomy, microbiology, and, nearer our concerns here, animal behavior.

training-for-the-unexpected hypothesis. The years-long project would yield both types of ethogram: one that organized and classified play behavior according to its resemblance to serious behavior, and another that distinguished behavior that was self-handicapping from that which was not. The plan was to observe groups of five species of monkey, giving particular attention to the animals' play and specifically to what was termed *play patterns*, defined as "recognizable behavioral unit[s] consisting of either a single coordinated movement or of a brief sequence of such movements."[10] No one would say the term possessed molecular precision, but in the unruly realm of monkey behavior it may have been all the precision that was possible.

For the first ethogram, two members of the team made video recordings of several groups of hanuman langurs in Bhangarh, Rajasthan, over three years. The team found that the langurs had forty-eight playful behavior patterns. Of these, sixteen—among them tumbling, somersaulting, and swinging—were, in the researchers' words, "totally dissimilar to any serious behavior known to us."[11] These were unlikely to be training for any practical activity. Nineteen of the forty-eight seemed part-serious, part-playful. Among these was a "play lunge," a quick, darting move like a play attack. It might seem serious, but since no fight ensued, it was fundamentally playful. The remaining thirteen patterns were identical to serious behavior. Clearly, the play patterns that partly or fully resembled serious behavior might serve as training for that behavior and would be explained by Groos's practice hypothesis or its more recent iterations.

To construct the other ethogram that would distinguish behavior that was self-handicapping, the researchers used the findings of the langur study and video recordings of four species of monkey living in zoos in the Czech Republic, Germany, and Switzerland. Each zoo provided the monkeys seminatural environments furnished with trees, ropes, and ledges—that is, ample opportunities for play. When Petrů's team reviewed the recordings, they identified seventy-four play patterns. Thirty-five, such as leaping and hopping in place, were used by all five species. The other thirty-nine were used by only some.

Why did some species use certain play patterns and not others? The likely answer, so the researchers thought, was owed to their habitats. Diana monkeys spent most of their time in trees and were strangers to the ground, so perhaps it was unsurprising that they didn't somersault; similarly, it was unsurprising that their more earthbound relatives—the langurs, de Brazza's monkeys, and vervet monkeys—did. Only the langurs and vervet monkeys deliberately jumped onto branches that yielded to their weight or made for an insecure perch, and only the de Brazza's monkeys grabbed branches and shook them by bouncing rhythmically with their whole bodies. The researchers conjectured that these might be explained similarly—that is, the monkeys simply taking advantage of the opportunities available: more yielding branches in the vervets' environment, and more shakable branches in the de Brazza's monkeys'.

But some behavior posed questions that were harder to answer. The monkeys in zoos had access to less space and fewer trees than did the free-ranging langurs, and so fewer obvious opportunities for play. Yet their play repertoire was wider. The researchers considered several possible reasons. Perhaps since the captive monkeys were fed regularly and so did not need to forage, they had more time to play and more time to invent new kinds of play. Or perhaps the langurs actually *were* playing as much or more than the captive monkeys, but doing it out of view of the cameras. In the end, though, the researchers agreed on a different answer. They knew that primates could find ways to play even in barren surroundings; in a well-known study, a chimpanzee had invented play patterns in an environment emptied of much to play with.[12] The team concluded that the captive monkeys, precisely because their environs offered fewer opportunities for play, were inspired to create new ones.

Parkour, Parkour! More Self-Handicapping Habits of Monkeys

All these findings, while intriguing, were tangents from the project's larger purpose: an ethogram that enumerated and described self-handicapping play. There turned out to be a great deal of self-handicapping play to enumerate and describe. One way the langurs self-handicapped was by making quick lateral movements of the head, thus deliberately compromising their vision and balance. The de Brazza's monkeys jumped on one another and tried to stay there for a few seconds, thus deliberately compromising their stability. When chased, the patas monkeys changed direction often and engaged in a bit of parkour—bouncing off walls—thus deliberately compromising their forward motion and lessening the chance for a real escape.

If the monkeys' self-handicapping play was impressive in its variety, it was just as impressive in its number. Of the seventy-four play patterns used by all five species, the researchers judged nearly half (thirty-three of those seventy-four) to be self-handicapping.

Natural Selection, Like Play,
Is Preparation for the Unexpected

Although in the short term the play of an individual animal is wasteful, exacting significant costs in that animal's time and energy, it is valuable and even necessary in the long term to prepare that animal for the unexpected. In this way it is quite like natural selection. Over time, existing threats to an animal such as predators and disease may increase, and new ones may emerge. Habitats, with their attendant resources, may diminish or disappear altogether. In the short term such changes can exact great costs, resulting in death of a great many individuals and even the extinction of an entire species. Yet in the long term, natural selection produces adaptations that enable an evolutionary line—much like an animal learning from play—to take advantage of such changes, to adjust to losses, to exploit opportunities, and generally to adapt to circumstances that cannot be foreseen or predicted.

As the research of Petrů's team demonstrates, the forms of play are complex and astonishingly varied. Marc Bekoff has called them a "behavioral kaleidoscope."[13] If ethologists such as Sharpe, Spinka, and Petrů are looking through the kaleidoscope's eyepiece, another group of scientists are working to understand the kaleidoscope itself—here meaning animals' brains and nervous systems. They conduct their studies not on sunbaked South African savannas or in mist-shrouded Scottish hills, but under the artificial lighting of basement laboratories in university science departments. Their tools are not binoculars and field notebooks, but mazes, microscopes, and chemical dyes. Their training, their very questions, belong to the fields of research known as *cognitive* and *behavioral neuroscience.*

Discoveries in these fields may have identified another similarity between play and natural selection, and yet another way in which life itself is playful.

CHAPTER 4

"Let's Go Tickle Some Rats":
The Neuroscience of Play

Like many scientists who study animal play, Jaak Panksepp was something of a rebel. When he was beginning his career in the late 1960s and early 1970s, two approaches to mental illness reigned. One was behavioral neuroscience, the study of physiological, genetic, and developmental mechanisms of behavior. The other was cognitive neuroscience, the study of the neural connections involved in the mental processes that underlie cognition. Neither approach had a place for emotions, and in those years many working in neuroscience considered emotions a subject unworthy of serious study.

Panksepp had reason to think otherwise. As an undergraduate, he had worked as an orderly in a psychiatric hospital, where he came to know patients afflicted with mental illnesses, some of them quite severe. In that work he began to appreciate how emotions, especially unbalanced ones, might wreak havoc on a person's life. Years later, as part of his work toward a PhD in clinical psychology, Panksepp mapped subcortical regions of rodent brains for a certain manifestation of "negative emotion"—aggressive behaviors. He grew convinced that the neurological networks that produce and channel those behaviors in humans were ancient, beginning perhaps as long ago as the first appearance of mammals. Emotions, he thought, were fundamental to mental development, and neuroscientists were making a serious mistake by disregarding them.

By the 1990s Panksepp was faculty at Bowling Green State University, where he devoted much of his research to establishing evidence for these views, pioneering the field of study he had named *affective neuroscience*. Among his ongoing experiments was an investigation of sounds made by laboratory rats, what he called "ultrasonically measured emotional vocalizations." They had a frequency of 50 kHz, well above the range detectable by the human ear. So that he and his students might hear them, Panksepp repurposed a bat detector, an instrument that converts bats' echolocation signals to frequencies audible to humans. When he placed the detector near rats that were playing, the noises he heard from its speaker were something like chirps. He suspected that they were the sound of rats having fun, and one morning he awoke wondering whether they were the rat counterpart of human laughter. If they were, he thought, there might be a way to be sure. When he arrived at his lab and met his research assistant, he said, "Jeff, let's go tickle some rats."[1] To maintain scientific rigor they considered using mechanical means, but decided the approach was too, er, mechanical, and simply used their hands. The rats they tickled came back for more, and the chirping of the colony increased dramatically. Panksepp said, "It sounded like a playground."* The experiment gained Panksepp brief notoriety worldwide as the "rat tickler," an identifier used in several of his many obituaries (he died in 2017).

The Nature of Rat Play

A rather playful sort himself, Panksepp once conducted an informal experiment by inviting members of various demographics to watch a video of two rats in a flurry of activity and to decide

---

* Animal laughter may be a subject deserving of its own book. Researchers have witnessed laughter in chimpanzees and Barbary macaques elicited by tickling and wrestling.

whether they were fighting or just enjoying rough-and-tumble play. One group, composed of his fellow professors, graduate students, and some higher-performing undergrads, called it fighting. Most from a group of undergraduates with slightly less distinguished academic records said it was play. A group of schoolchildren ages four to seven without hesitation called it play. The schoolchildren were right. Was this a demonstration of transference, projection, a bit of localized anthropomorphism? Or had the earnest and industrious professors and graduate students simply forgotten what play looks like—or since rat play is unlike human play—what it *might* look like?

Rat fighting is like a back-alley knife brawl. The participants may inflict serious harm. When they pounce, they will bite or try to bite the other's rump, flanks, or even face. They fight to win, attacking their opponents and defending themselves simultaneously. But rat play fighting is more like a pillow fight. A play-fighting rat will pounce, but it doesn't try to bite rump, flanks, or face. Instead, it tries to reach the nape but makes no attempt to injure; it only nuzzles with the tip of its snout or bites gently without breaking skin. It lunges in a way that leaves it vulnerable and open to the playful lunges of the other. If it makes defensive moves at all, it makes them more slowly. What's more, it and its opponent often reverse roles, switching from attacker to defender and back again.

Play-fighting rats follow a loose protocol. A play bout typically begins when one rat pounces from the rear. The other may then employ one of several strategies to protect the nape. It may simply run or leap away. It may adopt the "supine defense"—protecting the nape by falling backward or rolling onto its back. Or it may adopt the "standing defense": twisting around, standing on its hind legs, and using "hip-slams" while boxing with its forelegs—and so supplying us with an image that belies any conception of rats as unlovable. The standing defense may cause the attacking rat to lose its balance and expose its nape, turning its position of advantage into one of disadvantage. The nuzzler may thus become the nuzzled.

Practice for Sex

The term *play fighting* as applied to rats is misleading. Rat play fighting is not practice for fighting so much as it is practice for having sex. The basic move of play-fighting juveniles—to nuzzle or bite gently the other's nape—is also how a male rat of reproductive age with amorous intentions approaches a female rat of reproductive age.[*] Since the play-fighting juveniles are not yet of reproductive age, and since the recipient of their nuzzling is as likely to be male as female, it is a rehearsal—and much needed. One research review observes, "If deprived of peer-peer play as juveniles, rats, in adulthood, will perform poorly sexually."[2] And by "poorly" is meant very poorly. Play-deprived adult rats may try to mount the head of a receptive female.

Since the play fighting among juvenile rats is practice for sex—and therefore reproduction—it provides an adaptive advantage. That advantage cannot be fully realized until the rats are adults and able to reproduce. However, the behavior provides other advantages, too, some of which are immediate.

---

[*] Female rats solicit sexual approaches by wiggling their ears; by darting out in front of males they regulate the pace of mounts and copulations.

Reducing Stress

An animal that feels threatened becomes watchful and attentive. This behavior is its own adaptive advantage, and it may well be lifesaving. Yet a threatened animal feels stress, and when the threat is diminished or gone, the stress may linger and become a serious liability, impairing motor functions and reflexes.[3] Stress can also endanger an animal's long-term health because energies channeled to muscles by stress hormones are taken from the immune system.[4] When stress is unnecessary to an animal's survival, that animal has ample cause to be rid of it.

Rats rid themselves of stress in two ways. One is with social grooming: gently nibbling each other's fur. The other is by play fighting. In a number of studies, rats ceased play fighting when they were exposed to the scent of a cat; when the scent was removed, they resumed, in some cases with an even greater intensity. This "rebound effect" suggested the rats were play fighting not merely because they could, but because they were working to relieve stress caused by those feline pheromones.

Enhancing Social Competency

In the 1970s and 1980s scientists became interested in how the development of play influenced cognition and sexual behavior. Most notable among these scientists was Dorothy Einon, a lecturer in psychology at University College London. In 1977 she and colleague Michael J. Morgan hypothesized that if a rat was denied play as a juvenile, then upon reaching adulthood it might be less sociable than a rat that had played. To test the idea, they needed to isolate play from other behavior somehow, and that would be tricky. Juvenile rats experience a range of social contact—social investigation in the form of sniffing, mutual grooming, and huddling for warmth—and all these no doubt help them become sociable adults. How could Einon and Morgan allow those experiences and still disallow play? Their rather ingenious answer was to take advantage of a behavioral feature particular to rats. They knew that adult female rats engaged in the whole repertoire of social behavior, but played little. So they contrived an experiment in which juvenile rats were denied contact with rats that might play, but *were* allowed contact with adult females. This provided a juvenile period normal in every sense but one: it was absent play.

When the recreation-deprived rats reached adulthood, Einon and Morgan introduced them to a colony, where they proved to be something of a social disaster. The rats were unable to, as we might say, "read the room." They were not properly submissive in the presence of dominant males, exhibiting an attitude with, well, attitude—which prompted those dominant males to begin fights. And they overreacted to benign social contact such as sniffing, misjudging it as aggression and starting fights themselves.[5]

Einon and Morgan had shown that play fighting supplies rats with another benefit: "social competency." It should not be confused with "social bonding," that somewhat higher behavioral bar that Lynda Sharpe would suggest was unrelated to play, at least in meerkats. Social competency does not mean defending others or developing long-term relationships with them. It only means getting along with them and generally staying out of trouble. It's not what you do when you stand up for a friend who has been bullied—it's what you do when you make polite conversation at a dinner party you're not enjoying.

Practice in Defusing Conflict

Play fighting seems also to help rats refine a social skill that is both more sophisticated and specialized: a proficiency in defusing conflict. To understand how a rat might quell discord and friction, we'll first need to understand how those things arise. To do that, we'll need to learn something about rat society. Free-living rats populate a colony that has a dominant male who protects it. If he does not recognize a rat, he regards it as an intruder and attacks. The other rats, both male and female, so as not to be mistaken for intruders, groom the dominant rat often and engage him in play fighting occasionally—all this by way of saying, "No worries. It's just me again."

The dominant rat retains his social status by sometimes preventing other males from feeding and (less often) preventing them from mating. For this reason a subordinate male rat in a colony who wishes to eat regularly and mate occasionally would be prudent to stay in the dominant rat's good graces, and he can do so by initiating a playful attack now and then on the dominant rat's nape. When the dominant rat twists around and begins his playful counterattack, the subordinate, if he still wishes to stay in the dominant's good graces, responds with the supine defense. By falling or rolling onto his back, he says in effect, "I'm not a threat to your dominant status, and I very much still respect you." If the dominant is satisfied, and he likely is, he moves on to other matters of interest, and the subordinate rolls back onto his feet and also directs his attention elsewhere. The dominant rat has reestablished his dominance, the subordinate rat has recognized that dominance, and together they have brought about a de-escalation, shutting down a conflict before it could begin. No blood was shed, and the only thing hurt was the subordinate's self-esteem, if that. All is well in the republic of rats.

A Play-Fighting Puzzle

As a rat ages, its play-fighting choreography stays more or less the same; what changes are the uses of that choreography. Rats learn and refine the moves of play fighting during their juvenile period, when they are between twenty-three and sixty days of age. All those moves resemble sexual behavior and serve as practice for sex. By the time rats have passed through the juvenile period into adulthood, most of their fighting is real fighting—attacks on the flanks or face and inflicting injury. But adult rats don't abandon play fighting altogether. Their play fighting is somewhat rougher than the kind they engaged in as juveniles, the biting not quite as gentle. But it's clearly play: the moves and the target—the nape—are the same.

The play fighting bouts of rats are brief, often lasting less than a minute. So that the Pellises might understand exactly what happened in those bouts, they videotaped of them, played the tapes back at slow speeds, and paused them at critical points for closer study. In what is surely one of academia's more unusual examples of interdisciplinary appropriation, they documented the rats' moves with Laban movement analysis, a method choreographers use to describe human movement.[6] They found that as their subjects grew older, their movements became more forceful and controlled. They also found that certain moves, particularly some performed during the juvenile period, were quite perplexing.

While still weaning, rats play fight using the standing defense. The standing defense has an obvious advantage: it puts the defending rat in a good position to turn the tables and make offensive moves. So the Pellises were surprised to find that when the rats entered the juvenile period, they far more often used the supine defense—rolling onto their backs and protecting their napes. It's a position in which all four limbs are off the ground and no offensive move can be made, the very image of helplessness, hardly a defense at all.

That wasn't the only surprise. An adult rat typically pins its partner with its forelimbs, keeping its hind legs on the ground and thereby retaining some advantage of leverage. But the Pellises found that the juvenile rats pinned their opponents by sitting on them with all four limbs. A precarious position. And since the supine partner is not likely to stay supine, one that is probably fleeting.* Pinning with four limbs seemed all but certain to render a rat vulnerable to the other's attacks.

The Pellises' explanation for these moves closely aligned with ideas of Maxeen Biben described in the last chapter. A

---

* At the end of the juvenile period and the beginning of puberty, there was a clear split between the sexes. Females continued to use the supine defense, while the play-fighting strategy of males became a bit more complicated. The types of defense males used varied with their social status and with the social status of their partner. Dominant males used the standing defense with all partners. Subordinate males used the standing defense with females and other subordinate males, but used the supine defense with dominant males.

play-fighting rat is not interested so much in winning as in continuing the play fight. In using the supine defense, pinning with four limbs, and surrendering its advantage even as it may be about to win, the rat ensures that it does.

Training for the Unexpected, the Sequel

Spinka, Newberry, and Bekoff would opine that in allowing themselves to be thrown off-balance, the rats were giving their motor functions practice in recovering that balance, and learning to maintain control over those functions by keeping emotions in check. By putting themselves in disadvantageous positions, like the flop-over loving piglet, the rats were training for the unexpected.*

Which brings us, at long last, back to tickling. Since the subject has no single widely recognized expert, we are free to appeal to a general authority on behavior: Darwin. That one of the most formidable figures in the history of science should have ruminated on the nature of tickling may be surprising—until one appreciates his spirited relations with his children. Darwin ensured that Down House's opportunities for play included not only that portable wooden slide, but a rope swing hung over the first-floor landing. He encouraged his children to treat the gardens and surrounding countryside as an open-air laboratory and—since

---

* That the rats began to use self-defeating tactics during the juvenile period—during which rats play the most (at least an hour a day) and the period most important to their development as social beings—was significant. The finding provided another case in support of Byers and Walker's sensitive period hypothesis.

play and experimentation were hardly discrete endeavors—a playground. He mixed them freely himself, in one case applying a rather exacting proposition to a kind of play for which one imagines he had firsthand experience: "[A] child can hardly tickle itself," since "the precise point to be touched must not be known."[7] Most would agree. For the tickling to be effective, it must surprise. Humans being tickled are in disadvantageous, vulnerable positions. Those who allow and even enjoy being tickled may not know it, but they are learning what it feels like to lose control, and their memory of that experience is likely to prove valuable when they are confronted with a real threat. Being tickled is a kind of training for the unexpected.

Among play-fighting rats, tickling is better termed *nuzzling*. A rat being nuzzled makes defensive moves initially, but allows itself to be nuzzled and thus, like the humans allowing themselves to be tickled, train for the unexpected. But the nuzzled rat is not the only one who benefits from the interaction. The Pellises found that a rat who nuzzles gains as well. In a series of experiments, they drugged some members of a group of rats so they did not resist nuzzling and allowed all in the group to play. The rats who had not been drugged began to nuzzle the rats that had been drugged. The Pellises saw something they hadn't expected. When the nuzzling rats met no resistance from the drugged rats, they stopped. Then they did what rats often do when feeling anxious: they started to dig into the ground—or, in this case, into the litter in their chambers.[8] Clearly, they were dismayed that their nuzzling failed to have an effect. Since a rat is attentive to the response—or lack thereof—of the rat it is nuzzling, it is, in the very act of nuzzling, developing social competency.

For rats, then, play offers no fewer than five adaptive advantages: much-needed practice for sex, alleviating stress, defusing conflict, training for the unexpected, and developing and enhancing social competence.

The Four Questions of Nikolaas Tinbergen

Nikolaas "Niko" Tinbergen was a Dutch biologist and ornithologist who is regarded as one of the founders of modern ethology. He maintained that to fully understand a given behavior in an animal, you'd need to understand four aspects of that behavior: its adaptive advantage, its development throughout an individual animal's life, the physical features of the animal that make it possible, and, finally, its evolution in the species over time.[9] Since these aspects are deeply interrelated and explicable only by reference to the behavior as a whole, you'd also need to understand how each affected, and was affected by, the others. In interrogative form, the aspects came to be called "Tinbergen's four questions." Studies of animal play conducted in natural or seminatural habitats—such as Sharpe's—focused on the first: play's adaptive advantage. Studies conducted in laboratories—such as Einon's—focused on the second: its development throughout an individual animal's life. The Pellises, following Tinbergen's call for a holistic approach, worked to answer both questions, and to imagine how those answers might offer insight into the third: the physical features that make play possible. The physical features they paid most attention to were those shared by all animals that play: the nervous system and the brain.

The Brain, the Nervous System, and Play

The Pellises chose rats as their test subjects for the same reason that Panksepp did, and the same reason most neuroscientists do. They are small, easy to transport, and easy to maintain. Since thousands of studies have used rats, we know more about their anatomy and nervous system than those of any other mammal.* We also know more about their brains. Rats' brains are far smaller than ours (about the size and weight of a raisin), and far simpler. But since their major features are analogous to features of our own brains, and because their neural networks are much like ours, they serve as good models.

A quick refresher in things neurological. The brains of mammals, ours included, have three main parts. From front to back they are the *cerebrum*, the *cerebellum*, and the *brain stem*. The cerebrum, composed of right and left hemispheres, is the largest of these parts and performs many of the functions we call thinking. You're using your cerebrum to make sense of this sentence. The cerebellum, that smaller cauliflower-shaped organ tucked behind and a bit beneath the cerebrum, coordinates muscle movements and enables you to maintain balance. Unless you're reading this

---

* Because their genomes overlap with ours, they are valuable—even indispensable—as surrogates for humans. Mice, not rats, were used to test COVID-19 vaccines, but rats have long been used to test cancer drugs and HIV antiretrovirals. If you've ever received a flu vaccine, you may have a rat in one of the many laboratories associated with the Centers for Disease Control to thank for it.

Canada and many European nations have governing bodies that oversee research on rats and mice. The U.S. Animal Welfare Act protects most animals used for research, but exempts rats and mice, and rules for their treatment vary from institution to institution. Most offer a course in how to handle the animals in ways that minimize suffering and stress. Responding to a 2010 study showing that handling lab rats by the tail causes them anxiety, most labs found alternatives—such as lifting them with cupped hands.

on a bicycle or a tightrope, you're not using much of it now. The brain stem, true to its name, looks like a thick stem and performs the automatic functions we seldom think about, such as regulating body temperature and breathing. It sits behind and beneath the cerebellum, connecting the cerebellum and cerebrum to the spinal cord.

A single line drawn through the center of these three parts—*cerebrum* to *cerebellum* to *brain stem*—would be in the shape of a question mark. It's a particularly apt metaphor, since those three pounds of tissue represent mysteries wrapped in mysteries. Studies of the brain are rife with conflicting findings, provisional results, and muted equivocations such as "plays a role in" or "is associated with." Of the brain there is much we don't know.

But we do know that the brain mechanisms that enable or encourage play are remarkably specialized, and that specific networks in the nervous system and regions in the brain are dedicated to play. We also know something of the brain chemistry involved. A neurotransmitter is a chemical that carries signals between neurons and other cells throughout the body. Certain neurotransmitters seem particularly important at specific moments in a bout of play. Serotonin, for instance, must be nearly absent if play is to occur at all, and dopamine seems necessary for it to begin.

In 1992 Sergio and Vivien Pellis and graduate student Ian Whishaw attempted to identify parts of rats' brains specifically involved with play.[10] They began with a rather serious intervention—removing the entire cortices of newborn rats. In humans, the cortex is the folded grayish tissue that covers the cerebrum and is generally understood to be the source of planning, decision-making, and the other cognitive functions termed *executive.* Rat cortices are smoother than ours, with far fewer folds, but they provide the rat with those same functions—or, we may suppose, rudimentary versions of them. Since removing the cortex was a significant excision, it came as a surprise to all three researchers when, in the days following, the rats played like any rat with an intact brain. It came as a further surprise that when they reached the juvenile period, the time when rats with intact brains stop using the standing defense and begin to prefer the supine defense—falling or rolling over onto their backs—these rats continued to use the standing defense. The clear indication was that whatever triggered the switch to a supine defense in juveniles with intact brains was somewhere in the cortex.[*]

But where? The cortex has many areas, each with specialized functions. Sergio Pellis was particularly interested in finding out which area triggered the switch. He and a new team of students began by focusing on the *prefrontal cortex*, an area at the front of the skull that neuroscientists believe is responsible for several of the cortex's executive functions. They were curious about two parts of the prefrontal cortex in particular. One was the *orbital frontal cortex*, so called because in humans it lies just above the areas around the eyes that anatomists term orbits and is thought to be responsible for "context-specific responding"—that is, recognizing a given situation and acting appropriately.

The team made lesions in the orbital frontal cortices of a group of newborns. As expected, these had a dramatic effect on

---

[*] Upon finding no evidence that play in meerkats served as practice for social bonding, Lynda Sharpe inferred much the same, writing, "I conclude that the most likely function of play (based on play's ubiquitous characteristics, and the findings of neurological research on rats) is the promotion of growth of the cerebral cortex." Sharpe, "Play and Social Relationships," 1.

the rats' behavior. The team found that when the rats reached adulthood and were introduced to a colony, they readily engaged in play fights with the dominant rat, but when he began his playful counterattack, they did not fall into a supine position as would a normal subordinate. Instead, they responded with the standing defense. The dominant rat perceived this as the beginning of a real attack, and the play fight quickly became a real one.[11] It seemed that rats with lesions in their orbital frontal cortices could not recognize the social status of other rats.

Heather Bell, a doctoral candidate working under the Pellises, was particularly curious about the *medial prefrontal cortex*, the part of the brain lying just above the orbital frontal cortex and believed to be involved with social cognition—that is, recognizing others. In 2010 she and her team found that rats with lesions in their medial prefrontal cortices were less likely to initiate play fights, but when they did, they seemed confused, at some moments using the moves of play fights, at others using the moves of real fights.[12] The play-fighting partner, whether a dominant or another subordinate, regarded the latter as a signal that the fighting was no longer playful. Here again, matters escalated quickly with unpleasant results.

Both studies made clear that damage to the orbital frontal cortex and the medial prefrontal cortex produced social incompetence, the same kind as shown by rats denied play as juveniles. In rats here was evidence for a cause and effect between play and the development of parts of the brain's cortex. The question now became Which was the cause, and which was the effect? That is, did these parts of the brain enable play, or did play somehow develop these parts of the cortex? The answer, many suspected, might be found through a still-finer analysis.

Recall that John Byers had suspected that play and synaptic pruning were related. Bell was inspired to devise a new experiment that might show how. She divided newborn rats into three groups. Individuals from one group were each housed with an adult female. Since she didn't play, the newborns didn't either. Individuals from a second group were housed with one other newborn and thus ensured at least an hour of play every day. Individuals from a third group were housed with several other newborns. Since animals seeing other animals already at play want to join in (recall play's "contagion effect" and the author's wish to follow Jimmy off the bridge), individuals from this group played longer than the usual hour.

When rats from each group ended their juvenile period, Bell examined their prefrontal cortices. She found evidence of synaptic pruning in the cortices of the rats who had been allowed play, but no such evidence in the cortices of the rats who had been denied it. Clearly, neurological changes—changes on a fundamental, cellular scale—had been brought about by play.

Amygdalae and the Fear Circuit

Deep in regions of the cerebrum termed *subcortical* are two small, centimeter-wide, oval-shaped structures, one in each hemisphere. These are *amygdalae.* Amygdalae form and store emotional memories, especially the "negative" sort—that is, memories of experiences that aroused anxiety, fear, and aggression.

If the emotional memories created by an animal's amygdalae are of a real and recurring threat, they are useful and perhaps lifesaving. But if those memories are of events that are not threats or are threats no longer, they become a disadvantage, making an animal so fearful that it is unwilling or unable to engage in everyday activities. Suppose that, as you are reading this, you hear a loud noise—say a sharp crack, and it seems to have come from the kitchen. Your adorable but sometimes clumsy German shepherd has again knocked over a floor mop. You know it's nothing to be concerned about. But your amygdalae, with an emotional memory of a loud noise associated with a threat, send nerve impulses to your medial prefrontal cortex, which returns the call with other signals, creating a back-and-forth negotiation along a whole network of synapses that Roger Marek of Australia's Queensland Brain Institute calls a "fear circuit."

What has all this to do with play? Quite a lot, especially if the play is play fighting. By 2009, the Pellises knew that any understanding of rats' brains was far from complete, but they were convinced that play stimulates neural networks shared by the orbital frontal cortex, the medial prefrontal cortex, and the amygdalae. Specifically, play fighting made changes in neurons of the orbital frontal cortex and the medial prefrontal cortex that enabled rats to recognize the social status of potential playmates, and to differentiate between tactics appropriate to play fighting and tactics appropriate to actual fighting. Moreover, both cortices suppressed the amygdalae. Without their intervention, the amygdalae might cause a rat to be either so afraid that it wouldn't play at all, or so aggressive that it would make others too afraid to play with it.

But for a juvenile rat with an intact brain, those two cortices developed and so enabled a negotiation with the amygdalae that improved the rat's ability to recognize the status of potential playmates and to discriminate among play-fighting tactics. That negotiation might be difficult to visualize. So let's indulge in a bit of anthropomorphism, not of play-fighting rats, but of the brain of one play-fighting rat during a moment of uncertainty.

The Brain at Play—a Dramatization

Imagine a darkened theater. Sounds of confusion. A struggle.

AMYGDALAE: "That hurt. Let's bite back."

MEDIAL PREFRONTAL CORTEX: "Just a sec. I think he didn't mean it."

ORBITAL FRONTAL CORTEX: "That bite was from the dominant male. Better go to supine, quick."

AMYGDALAE: "Bite back."

MEDIAL PREFRONTAL CORTEX: "Lemme think. Okay. Two things. First: so far he's been playing. He's had no reason to begin fighting now. Second—"

AMYGDALAE (*interrupting*): "Bite back."

MEDIAL PREFRONTAL CORTEX: "—*ahem*. Second: last time this happened it was a mistake."

AMYGDALAE: "Bite back."

ORBITAL FRONTAL CORTEX (*to Amygdala*): "What part of *dominant adult male* don't you understand?"

AMYGDALAE: "Bite—"

MEDIAL PREFRONTAL CORTEX (*interrupting*): "Overruled. I'm in total agreement with orbital frontal cortex. We're going supine."

HYPOTHALAMUS (*after a brief pause and to no one in particular*):
"Well. That was fun."*

This back-and-forth between the two cortical regions and
the amygdalae can occur only during play fighting. Every time
a rat begins to play fight, the connections are reinforced and
strengthened, and its play-fighting abilities are ratcheted up. In
a feedback loop, the brain refines play fighting, and play fighting
refines the brain. What had been only a correlation—that is,
more play correlates with more neurological development—was
shown to be a traceable cause and effect, and a reciprocal one
at that. The Pellises had found that in laboratory rats the fight-
or-flight response was hereditary, but like an electrical switch,
it came set either to *on* or *off*. To become well-adjusted, a rat
would need practice in discriminating between situations in
which the better response is fight and situations in which the
better response is flight. It would need to toggle that switch.
And play—particularly the rough-and-tumble variety—provided
opportunities for that toggling.

* In several ways (other than regions of rats' brains voicing English), the dramatization
is unrealistic. First, while it might take half a minute to read aloud, since nerve impulses
travel as fast as 275 miles an hour, the entire negotiation would have been completed
in half a second. Second, it's an oversimplification. As neuroscientists study the brain's
regions on ever-finer scales, they have found remarkable specialization, but remarkable redundancies as well. And certain important functions are shared. For instance,
neuroscientists are increasingly persuaded that the cerebellum is involved with and
responsible for some of the brain's "higher" functions—such as interpreting sights and
sounds—long believed to be the province of the cerebrum alone. A structure in the
upper brain stem called the *striatum* may intercede in activities also long attributed to
the medial prefrontal cortex. In 1993 the Pellises and Mario McKenna found that a rat
whose striatum had been compromised will still play fight, but will confuse offensive
moves with defensive moves and sometimes confuse them midmove, for instance, lunging
at a playmate's nape and before getting there twisting away, as though the playmate had
tried to lunge at *its* nape. Pellis et al., "Some Subordinates Are More Equal."

"The Decline of Play and the Rise of Psychopathology in Children and Adolescents"

The Pellises maintain that although rats may serve as model animals, to apply their behavior and neurology indiscriminately to other species and classes of animals is likely unwise. Nonetheless, in *The Playful Brain*, they did allow themselves to suppose that rough-and-tumble play encourages the development of social skills in humans, too—and that it may be necessary to our psychological and emotional health.

A researcher who arrives at a similar conclusion is Stuart Brown, founder of the National Institute for Play. A clinical psychologist by training, he came upon the subject of play more or less accidentally. In the 1960s he was part of three research teams, each working on a separate study. One team was examining the background of the Texas-tower mass murderer Charles Whitman, another the backgrounds of certain convicted murderers, and still another the backgrounds of a set of automobile drivers whose recklessness had caused fatalities. Brown notes, "What struck our separate research teams as unexpected was that . . . normal play behavior was virtually absent throughout the lives of highly violent, anti-social men regardless of demography."[13] All work and no play may make Jack a dull boy. It seems that it also makes him a dangerous one. Brown cautions that the "linkages from the objective findings in animal play deprivation to the clinical findings in humans are as yet unproven." But he allows, "The subcortical physiology and anatomy is similar, and the inability of play deprived animals to deter aggression, or to socialize comfortably with fellow pack members, is demonstrable."[14] Professor of psychology Peter Gray draws the same conclusion in a much-cited article with the no-nonsense title "The Decline of Play and the Rise of Psychopathology in Children and Adolescents."[15]

Readers of a certain age recall whole afternoons spent outside, in woods, in fields, in back lots or city neighborhoods, with no adult supervision whatsoever. If you are of that age, it's a memory of a time so unlike the present that you might wonder if you've been romanticizing a childhood that wasn't. But a review of research on play from the last two decades shows that the change in behavior was dramatic and well-documented.[16] Much has been written of what's been called nature-deficit disorder, and the benefits of unstructured and unsupervised play. In this regard Panksepp made an intriguing observation: "No one has yet explicitly conducted a play-deprivation study in our species, even though I do suspect we are currently in an unplanned cultural experiment of that kind. Too many youngsters of our species never get sufficient amounts of natural, self-generated play. If so, that may be one cause of our current epidemic of hyperkinetic kids with inadequate control over their own impulses."[17]

The Rough-and-Tumble Childhood Play of Ethologists

Many of the scientists mentioned here have written or spoken of their own childhood play. In each case, that play was unsupervised and resulted in bruises, cuts, and other unpleasantries. Lynda Sharpe recalls with characteristic humor, "My front teeth suffered irrevocably in the great roller-skating catastrophe of 1972."[18] Jaak Panksepp recalled childhood play in environs that would seem profoundly unsuited to it. In 1944, as the Red Army advanced upon Estonia, the Panksepps escaped to northern Germany, where for several months they lived in a displaced persons camp, and Jaak and his companions were afforded an uncommon playground.

> Once, about a quarter of a mile from camp—I think that was when we were in Merbeck, Germany—we were playing hide-and-seek and king-of-the-mountain in a dump for ruined war hardware, including the skeletal remains of German tanks and trucks. I took a big tumble off the top of a tank and fell on some rubble that skinned my head pretty badly. After I regained my wits, my brother helped me get home, but I was bleeding and wailing all the way.[19]

Researchers have given little attention to a specific kind of rough-and-tumble play: play fighting. The Pellises observe that many recent textbooks on child development neglect the behavior, despite that it may represent nearly 20 percent of spontaneous play in school playgrounds, it seems remarkably similar across cultures, and, so far as anyone can judge, it has changed little throughout history. Most significant to our interests here, it is the one kind of human play that most resembles the play of animals.[20]

Several of the ethologists we've met had firsthand experience with play fighting. Sergio Pellis participated in playground fights as a child in secondary school in Australia.[21] Panksepp recalled, "At one of our next camps, in Oldenburg, where the Estonian sector and the Latvian sector were divided by a soccer field, youngsters would gather on each side of the field holding clumps of grass—still weighted by dirt-clenching roots—as grenade-like ammunition and then start a *Lord of the Flies* sort of battle."[22]

Although rough-and-tumble play can cause injury, it may endow the brain with a means to keep emotions in check. Play fighting in particular may provide training for the unexpected, and necessary practice in social skills. Children denied the opportunity to engage in play fighting may become adults deficient in the ability to empathize, with little skill in negotiation and no notion of ambiguity. One can't help but wonder, Is it possible that some members of this generation of adults, politically polarized, with no ability to listen, let alone compromise, are this way because they did not play fight as children?[23]

The Surplus Resource Theory of Play

In 1984 Burghardt put forth a theory of play that addressed its development in the life of an individual animal. He called it the *surplus resource theory.*[24] It was an elaboration and extension of Spencer's surplus energy hypothesis, but where the only surplus Spencer accounted for was energy, Burghardt added others: an ample and reliable food source, ample protection from predators and extreme weather, and ample time to play.

The surplus resource theory accommodates the findings of the Pellises and others, but it does not focus on any particular animal, nor on any particular neurological mechanism or feature. Rather, it focuses on an animal's behavior and, from observations of that behavior, defines stages through which the animal's play develops. The first stage is *primary process play.* It's the repurposing of an excess, the fidgeting, squirming, and wriggling produced by "excess metabolic energy." Primary process play doesn't benefit the animal, but neither does it do harm. If for a minute or so you watch a young animal engaged in such play, you might think its movements look a lot like those of a captive or unhealthy animal performing stereotyped behavior. But watch for a while longer and you see that its movements are varied. That variability is one sign that the animal is healthy. Burghardt's theory posits that for some animals, primary process play may develop into *secondary process play*, which maintains or helps maintain the animal's physical and neurological condition, enhancing an animal's connective tissue, nervous system, or cardiovascular system. For some animals, the behaviors of secondary process play may develop into *tertiary process play*, the sort we've been most concerned with in this book. Tertiary process play provides practice for behaviors that might include sex, fighting, and defusing conflict, but whose range and kind are, for all intents and purposes, unbounded.

Animals engaging in secondary process play may continue to engage in primary process play, and animals engaging in tertiary process play may continue to engage in both secondary process play and primary process play. A human animal at play—say a chess player as she contemplates a move—might be partaking of tertiary process play by considering the merits of the Sicilian Defense, partaking of secondary process play by exercising neural processing, and partaking of primary process play by drumming her fingers.

Natural Selection, Like Play, Brings about Order

In his surplus resource theory, Burghardt posits that by organizing, channeling, and refining behavior through stages, the play of any animal brings order to that behavior. The Pellises and other researchers amassed a great deal of evidence showing that although the protocols of play vary from species to species, all give form to impulses and reflexes that would otherwise be erratic and unpredictable. The Pellises demonstrated that play also brings order to the brain and nervous system, enhancing neural networks and deepening connections among three otherwise loosely related regions in the brain—the orbital frontal cortex, the medial frontal cortex, and the amygdalae.

Physicists use the term *closed system* to describe a region that is isolated from its surroundings by a boundary across which matter or energy cannot pass. They call the tendency toward disorder and randomness *entropy*. One formulation of the second law of thermodynamics is that the entropy in a closed system can never decrease and can only increase. Since the universe as traditionally defined has nothing outside it, it is a closed system. Thus the universe on the whole tends toward entropy—disorder and randomness. But natural selection, in evolving living organisms, operates against this tendency. A living organism is an open system and can take energy from other places also within the universe and put that energy to its own uses. A living organism is an eddy or countercurrent, a small-scale—and momentary—reversal of the universe's ever-increasing entropy.

So to the list of features shared by play and natural selection, we may add another. As play brings order to an animal's behavior and to its brain and nervous system, so natural selection, in evolving living organisms, brings order to the universe. Play is natural selection writ small. Natural selection is play writ large.

CHAPTER 5

Courtly Canines:
Competing to Cooperate
and Cooperating to Compete

S ome fifteen thousand years ago, in the dense forests of what is now northern Europe, a wolf pack lived near a human settlement and regularly scavenged scraps of food left at its edges. The humans were wary of the wolves they heard howling at night but left them alone. The wolves had little use for the humans except as occasional providers of food.

But then something changed. Exactly what we can't know, but it may have been something like this. One of the humans, perhaps a young girl gathering mushrooms, spotted a wolf alone in a clearing and stopped to watch. The wolf, seeing her, likewise became still. Then the wolf, bolder and friendlier than others in its pack, approached her. The girl, now more curious than afraid, took a bit of food from her sack and tossed it toward the wolf. The wolf sniffed it, ate, and came nearer.

In time, the wolf was allowed inside the settlement. A few more wolves, seeing that the first was treated well, approached the camp and were also adopted by the humans. Soon one of the females bore cubs. The cubs were tamed and domesticated, and in a surprisingly short time—only forty or so generations—the wolves' appearance changed, with more and more having splotchy coats, floppy ears, and wagging tails. Just as important, their cognitive abilities evolved: they began to understand human gestures such as pointing. They could no longer be called wolves. They were dogs.[1]

Dogs' abilities to "read" humans combined with their hunting prowess made them especially valued. In the northern forests, humans' mobility and range of vision was limited. But in those same forests, dogs could chase, track, sniff out, and retrieve prey. Sharing territorial boundaries with human encampments, they could also warn of animal predators and human intruders. The wolves-turned-dogs gained at least as much by the arrangement, as they would be well-fed and cared for. The result is what Brian Hare, director of the Duke Canine Cognition Center at Duke University, calls "the most successful interspecific cooperative-communicative relationship in mammalian evolutionary history."[2]

Brian Hare has a mop of dark hair and an agreeable demeanor many would call boyish. In the late 1990s he was an undergraduate at Emory University in Atlanta and working as part of a research team headed by comparative psychologist Michael Tomasello. At the time, the mental capacity of chimpanzees was being compared to that of human toddlers, and Tomasello and his team were testing and refining that comparison. One finding in particular surprised them. When a researcher pointed to a cup, the toddlers looked to the cup; the chimps did not. Tomasello took this as evidence that the ability to follow pointing cues was a skill evolved only in humans. But Hare was less certain. As he recalled, "I told Mike, 'Um, I think my dog can do that.'"[3]

Hare became so intrigued by dogs that he made them his field of specialization. He is interested in many aspects of canine behavior, one of which is the role play may have figured in the evolution of wolves to dogs. Over generations, the wolves-turned-dogs became *prosocial*, a term animal behaviorists use to describe behaviors of tolerating others, cooperating well, and working to resolve conflicts.

Humans are also a prosocial species. We've alluded to studies suggesting how play—and a lack of it—might affect human sociability. Is it possible that just as we domesticated wolves, those wolves (or their dog descendants) helped to domesticate us? Hare thinks so, observing that while a wolf faced with a task such as lifting a door latch will try to solve it alone, a dog will look to you for help. In all likelihood, you—as that domesticated human—will give that help and lift the latch.

This mutual domestication may have come about, at least in part, through eye contact. A fascinating 2015 study showed that dogs' "gazing behavior" increased oxytocin levels in their human owners' urine. The effect was reciprocal. As the dogs' gazing made their owners behave more affectionately toward them, that affection in turn resulted in increased oxytocin concentrations in the dogs' urine. The study concluded that an "interspecies oxytocin-mediated positive loop may have supported the coevolution of human-dog bonding."[4] So, when some humans and dogs make eye contact for even a short period, they produce physiological changes in each other, the same kind of physiological changes that for many millennia have allowed us to live together, help each other, and—of course—play with each other.

The Canine Play Bow

Perhaps the best-known scholar of play is Dutch historian Johan Huizinga.* In his 1938 book *Homo Ludens* ("playing man"), he noted of dogs, "They invite one another to play by a certain

---

* Huizinga made a case that in essence animal play is no different from our own, writing, "All the basic factors of play, both individual and communal, are already present in animal life—to wit, contests, performances, exhibitions, challenges, preenings, struttings and showings-off, pretences and binding rules." Huizinga, *Homo Ludens*, 47.

ceremoniousness of attitude and gesture."[5] He was referencing a fascinating and much-studied act in the repertoire of canine behavior: the *play bow*. A dog bends its forelimbs so that its head and shoulders are lower than its hindquarters and may bark or wag its tail. Marc Bekoff called it "a highly ritualized and stereotyped movement that seems to function to stimulate recipients to engage (or continue to engage) in social play."[6] Since puppies play bow without training, the behavior seems instinctive. But because it's instinctive does not mean it isn't sophisticated. The behavior is made possible by a *theory of mind*—that is, the ability of an individual to correctly identify the mental state of another, to explain its current and predict its future behavior. Only through a canine theory of mind can the play bow work at all. A dog who sees another dog perform a play bow will understand that the other invites him to play. When the dog who performs the first bow sees the other bowing in return, he knows that his invitation has been accepted. The other dog trusts that the bow means that whatever follows—even behavior that looks and sounds like real fighting, such as growling or baring teeth—is still play.

A play bow is one of a range of signals that animals use to invite play and prevent it from turning aggressive. Voles, for instance, produce a particular pheromone; dwarf mongooses make distinctive vocalizations; and primates use a "play face," the mouth relaxed and slightly open.[7]

Keeping Play Fair

The kind of play that most often ensues after the canine play bow is play fighting—pouncing, wrestling, and biting gently without breaking skin. Dogs that play fight abide by certain rules of engagement. Like rats, they work to keep things fair. They don't bite as hard as they might, and they often reverse roles. A dog with the advantage of position might suddenly roll over on its back. Since play fights are spirited and energetic, on occasion one bites harder than it intended. Afterward it is likely to retreat and perform another play bow—saying in effect, "I'm sorry. I ask you to forgive me and hope that we can continue playing." In most cases, forgiveness is bestowed, and the play begins again.[8] Members of other play-fighting species keep things fair in other ways. Degus are small gerbil-like mammals native to central Chile. A degu who gains an advantage in a play fight, knocking his opponent over, does not press that advantage. He allows his opponent to regain his footing before reengaging.

To begin a play fight, dogs, rats, and degus must agree on a protocol; to continue, they must adhere to it. While animals that are play fighting are competing, they are also cooperating. Play employs both competition and cooperation and holds them in a dynamic equilibrium. In this way, too, it is like natural selection.

Natural Selection, like Play, Holds Competition and Collaboration in Dynamic Equilibrium

To understand how, we'll need to delve into the history of the theory—specifically, a refinement that was introduced and developed in the first decade of the twentieth century. Darwin acknowledged cases in which two organisms evolved in ways that each benefits the other, but they were not representative of the large body of examples from which he drew.* The "struggle for existence" as he first presented it, and as many of his advocates *re*presented it, was by and large a struggle against others. This emphasis was perhaps owed to the kinds of organisms Darwin observed during the nearly five-year voyage of the *Beagle*. The subjects that inspired his theory lived in the tropics, which have, by comparison with higher latitudes, more species in a given area, but fewer individuals representing any one species, and thus on the whole more competition than cooperation. A naturalist wishing to observe animals in large groups, and thus cooperation among individuals within those groups, might look elsewhere— say the skies above New England.

On an October afternoon many years ago, my high school soccer team was running drills. A formation of Canada geese came out of the north and flew overhead in that V formation that ornithologists call an echelon. Only a few of us noticed at first. Then another, larger formation followed, and another and another, until lines of geese—hundreds and hundreds of birds—were stretching across the whole sky. It was one of those moments when nature, for many of us perhaps too often a backdrop, simply would not be ignored. Our coach, shouting to be heard over the honking, called us into a huddle and, to our surprise, delivered an impromptu lesson in avian aerodynamics. He asked us to watch the geese in one leg of the echelon directly above us. He told us that each goose was getting a lift from the vortex created by the goose in front of it, that the lead position

---

* In *The Descent of Man*, Darwin notes (p. 162) that the "social qualities" of humans are necessary to the survival of both individuals and groups.

at the apex of the echelon was the most tiring, and that sooner or later the goose in that position would fall back along the column, and another would take its place. By this energy-saving strategy the geese might cover a thousand miles in a day. We watched until the last stragglers were almost out of sight and the honking had all but died away. Our coach looked at each of us for a moment and, in as serious a tone as I'd ever heard him use, said, "Gentlemen, that's teamwork."

Such cooperation was not among the behaviors described in *Origin* but was central to the interests of a man whom we might imagine as Darwin's intellectual doppelgänger: Russian naturalist Pyotr Kropotkin.

Kropotkin had a career so varied that it defies categorization. He was a revolutionary, an economic theoretician, a geographer whose work resulted in some of the first accurate maps of the physical features of Asia, a foremost theorist of the anarchist movement, for two years a prisoner held for subversive political activity, and for forty-one years a political exile. To all these one might add naturalist and evolutionary theorist. From 1862 to 1867, Kropotkin served as an army officer and a member of state-sponsored geographical expeditions charged with mapping the vast terrains of Siberia and northeast Asia. During spare moments on those expeditions, he studied flocks of birds and herds of fallow deer and wild horses, gradually coming to believe these species were on balance "the most numerous and the most prosperous" precisely because they were cooperative. In his 1902 book *Mutual Aid: A Factor of Evolution*, Kropotkin derived an interpretation of Darwin's theory that greatly enlarged its explanatory power. While natural selection certainly selected for better competitors, said Kropotkin, it also selected for better cooperators.[*]

In the twentieth century scientists confirmed Kropotkin's hypothesis, discovering all manner of cooperation not only between members of the same species, but between members of different species, a relationship called *symbiosis*. Some were *mutualistic*, in which both individuals benefit. Some were *commensal*, in which one benefits and the other does not, but neither is harmed. Some were *obligate*, in which each depends on the other for survival. And some were *facultative*, in which one or both might benefit, but either or both can also manage fairly well alone.

In the first half of the twentieth century, several biologists proposed that such cooperation operated at the cellular level and had been doing so for quite some time. They proposed that roughly 1.6 billion years ago, the organelles within a cell known as *mitochondria* were free-living, oxygen-respiring bacteria making do in a harsh world. Then one or several found refuge in the warm, wet, pH-balanced interior of a cell. In time the guest, perhaps uninvited but not entirely unwelcome, provided the cell with energy and disposed of its waste. The arrangement worked well; it was enhanced over time, and today the interdependence is so complete that the cells in your body would die without the mitochondria inside them.

In 1967 American biologist Lynn Margulis put forth a theory that other organelles, plastids and basal bodies, had also once

---

[*] Many have noted that since Charles Darwin spent the better part of his life in England, a nation that espoused free markets and thus competition, his theory may also have been influenced by human behavior in the realm of economics. Kropotkin's biographers note that his communist ideology led him to regard cooperation as an important factor in natural selection, and that his observations of animal cooperation in turn offered support for his communist ideology.

been free living, but had entered prokaryotic cells—simple cells without nuclei—and in time enabled their evolution into eukaryotic cells, that is, complex cells with nuclei. The relationship that resulted was called *endosymbiosis*, a symbiosis in which one organism lives inside another. In the years and decades that followed, experimental data from electron microscopy, genetics, and molecular biology provided evidence in support of the theory, and by the 1990s biologists recognized endosymbiosis among many organisms. Still more recently, evolutionary biologists have theorized that 4 billion years ago, before the evolution of DNA and proteins and cells, all life on earth was composed of much-simpler structures that had begun to work together. It seems that organisms' propensity to cooperate is both ancient and fundamental.

Natural selection evolves cooperation among organisms, cooperation produces order, and order produces expectations that it will continue. But natural selection also evolves organisms that subvert and undermine that order and exploit those expectations. Female fireflies of the genus *Photuris* emit light signals that females of another genus use to attract males from that genus. They approach the females, only to be captured and eaten. Such behaviors are not uncommon. Using mimicry and availing themselves of the advantages it provides, many organisms deceive their prey, their predators, and their rivals.

In a previous chapter we saw that play fighting enables rats to get along with others, stay out of trouble, and defuse conflict. These behaviors are reactive—that is, responses to a given situation. But play fighting also enables animals to develop skills that are proactive, skills that enable them to exploit the ambiguity in moments of play fighting, making it possible for them not merely to *respond* to a situation, but to *control* it. Play—and play fighting specifically—offers animals opportunities to deceive, and opportunities to learn how to deceive. Not all animals who play, play fairly.

Play and Deception—the Uses of Ambiguity

The rules of play fighting for any species may be well-defined and ordered, yet any play fighting may have moments during which one or both participants are uncertain of the other's intentions. These moments give both animals opportunities to practice a theory of mind, to negotiate and to develop skills in general social competence and social assessment, skills necessary to forestall the escalation of actual fights and to avoid them to begin with. But an individual animal might exploit those moments, using them to its own advantage in a way that might seem a bit wicked.

Any play fight inevitably has moments when one animal's move—or more precisely the motive behind that move—is unclear to the other. Say one animal in a play fight bites too hard. The bitten gives the biter the benefit of the doubt and continues playing. The biter bites again, and again too hard. The bitten thinks the biter may have stopped playing and is actually fighting, but can't be sure. And because it can't be sure, it is at a disadvantage. But the biter knows full well that it is biting too hard; it is deliberately pushing the limits of play for a reason.

Exactly what reason? Suppose that you are a subordinate rat living in a colony. Suppose further that you are not just any subordinate rat. You are a rat with resolve, an animal with ambition, a subordinate who doesn't wish to stay subordinate. Challenging the dominant rat directly is an all-or-nothing bet. If you win a fight with him, you become the dominant rat; if you lose a fight with him, you stay subordinate, with scars to show for your troubles. But a low-risk, fail-safe way to test your prospects is to initiate a play fight—not a real fight—with the dominant rat. Then do something that skirts the line into real fighting. Nibble a bit too hard, for instance. If the dominant rat makes a move that also verges on real fighting—say, attacking your flank—you can simply back off and do something a subordinate would normally do; for instance, roll over on your back. It would be saying, "Oops. My bad. Didn't mean to bite that hard." The dominant rat, believing your harder-than-usual bite was accidental, will probably also revert to behavior that is unequivocally playful.

Suppose though that when you deliberately nibble too hard, the dominant rat does *not* attack your flank, does *not* respond with an action that verges on real fighting. In that case perhaps he is genuinely frightened. If he is, then you have found his vulnerability and an opportunity to advance your social status. "We believe that this gray zone of uncertainty endows play fighting with its value as a tool for social assessment and manipulation," write the Pellises. "Taking advantage of the situation in a playful context . . . can be informative about the weakness of one's social partner."[9]

Play and Pretense among Macaques

Do other animals also use play—or rather, the ambiguity of play—to negotiate, assess, and manipulate? The answer, at least in the case of the old-world monkeys called macaques, seems to be yes. Macaques are a genus of twenty-two species for which the word *myriad* barely serves. Among them are the Barbary apes that scramble up and down Gibraltar, the white-haired elfin-like toque macaques of Sri Lanka, and the stump-tailed macaques, memorable for their red eye patches and leonine, heart-shaped coiffures, of south China. Macaques' appearances and chosen habitats differ greatly, but their social lives are much alike. Most macaque species live in groups of twenty to forty individuals, each group maintaining not one, but two dominance hierarchies. One is composed of females—grandmothers, mothers, and aunts, who care for the young and stay within the same group for their entire lives. Another is composed of males, each of whom joined the group during their adolescence after leaving the group into which they had been born.

The rhesus species of macaques are the most studied. Their hierarchies are rigid; the ranks of individuals are largely fixed, and individuals change ranks only rarely. For many years, ethologists assumed that the hierarchies of all macaque species were just as inflexible. They were perhaps mistaking the part for the whole. In the late 1990s Bernard Thierry of Université Louis-Pasteur, collected observations of seventeen species and found, to the surprise of many, that some of those hierarchies were rather pliant.[10] He was able to place all macaque hierarchies at points—or rather, since definitions of behavior are imprecise, regions—along a scale of tolerance. At one extreme you'd have Japanese macaques, the much-photographed "snow monkeys," often found soaking in hot springs, faces a pinkish red, hair spiked and wet. Recent research suggests they use the baths to relieve stress.[11] At least for a subordinate, there may be much stress to relieve. Japanese macaque hierarchies are so rigid that Thierry termed them "despotic." A subordinate will not take food from a dominant, but a dominant can—and often will—take food from a subordinate or strike it.

By way of contrast and at the other extreme, are Tonkean macaques, one of several species inhabiting the mountainous island of Sulawesi in the Malay Archipelago. Their gray faces and heavy brows, recalling portraits of somber British magistrates, hardly bespeak tolerance, much less indulgence. Yet their hierarchies, especially when compared to those of Japanese macaques, seem almost laissez-faire. A dominant Tonkean macaque whose food is taken by a subordinate might not bother to take it back. Naturally, such tolerance allows considerable social mobility. Like that rat with ambition, a subordinate Tonkean macaque wishing to change its status may challenge a dominant. It may, for instance, strike a dominant with a blow that might be felt as too forceful for play. If the dominant does not strike back, the subordinate may see an opportunity to advance. But if the dominant *does* strike back, and strikes with sufficient force, then play provides the subordinate a ready means of retreat. Although it meant the strike as an actual challenge, it may signal that it was not one and was only play.

Such an interchange requires considerable mental effort on the part of both animals. Each must judge whether the other wishes to begin a play fight or an actual fight, taking into consideration the forcefulness of the other's blow, the context of the interaction, the past behavior of the other, and the very character of the other—and the animals must do all this in seconds. Obviously, this requires a good deal of cognitive exertion, and it's easy to see why Japanese and rhesus macaques would conclude that such effort is not worth the social mobility it might afford. Sometimes life is easier when all just know their place and stay there.

Rats seem to develop play-fighting skills—along with the ability to use them to negotiate, assess, and manipulate—when they are juveniles. Is the juvenile period also important for the play of Tonkean macaques? Perhaps. Juveniles of all macaque species play fight, but only juvenile Tonkean macaques vary their targets and tactics.[12] In so doing both participants receive and deliver the unexpected, the sort of unexpected experienced by adult animals—that is, the behavior of others. The innovative play of juvenile Tonkean macaques seems to provide an education in that behavior that, upon reaching adulthood, they employ to their advantage. Adult Tonkean macaques use play to negotiate, assess, and manipulate.

An Example Nearer Home

What of yet another species of primate—us? Do we use play to negotiate, assess, and manipulate? The answer is a qualified yes, in some types of play in certain groups, and perhaps especially in verbal play among members of subcultures acquainted with actual fighting.[*] A vivid albeit fictional example is a scene from Martin Scorsese's 1990 film, *Goodfellas*. The animal in this case is a human, a mob boss named Tommy DeVito, ably played

---

[*] In parts of Ireland, an unusually harsh form of verbal play fighting is called "slagging."

by Joe Pesci. DeVito hangers-on are seated around a table at a restaurant, and DeVito tells a story that ends with a vulgar joke. Obliging laughter ensues, and one listener, a recent initiate with uncertain status within the group, says, "You're funny."

DeVito responds, "I'm funny how? Funny, like I'm a clown? Do I amuse you?"

The initiate laughs again, a bit confused. "I dunno. You're just funny."

DeVito persists, his tone now shading into threat. "How am I funny?"

Others at the table know DeVito's reputation for violence; they intercede and try to calm him. But he leans forward. "No. He said I'm funny. I wanna know how."

For a moment the initiate is afraid. Then, suddenly, he realizes that DeVito is provoking him, and he laughs, and laughter breaks around the table, along with palpable relief.

An ethologist viewing the scene might observe that DeVito is a dominant primate and not seeking to change his status. Yet a dominant's status is never secure; it may be challenged at any time, and a dominant wishing to keep his status must work to maintain and preserve it. That ethologist might say that DeVito has maintained and preserved his status by exploiting play's ambiguity. But he's also gained a bit of knowledge for future use. He has seen the initiate frightened and now knows what frightens him.

The initiate and others laughed because they felt relief. But what exactly is relief? Among other things, it's the pleasure of regaining control after a moment of disorientation and fear. So one reason the initiate and others laughed—the answer that Spinka and company might proffer—is that they felt the same pleasure.

There is a pleasure in the momentary disorientation of being tricked, or perhaps more precisely, a pleasure in learning that one has been tricked, that one has been, in a word, "played." It's a pleasure not confined to humans. A popular online video shows an orangutan watching a performance of the cup-and-ball trick. It watches the movement of ball and cups intently, until the performance's conclusion, when it is shown that the cup it thought would hold a ball is empty. Staring into the cup, confirming that against expectations that it *is* empty, the orangutan rolls onto its back with obvious delight. Because magic tricks are typically performed in an unthreatening context, the disorientation they offer is entirely benign. They are thus a means to train for the unexpected minus the fear and without the risk.

Canine Play and Ethics

Over the past several decades, the list of features thought to distinguish humans from animals has grown shorter and shorter. Morality may be the last item remaining on that list, and its inclusion seems increasingly tenuous. Ethologists have identified what seems to be moral behavior in many animals,[13] documenting cases of what many humans have long known: dogs have a sense of fairness, a willingness to comfort another suffering injury, and a readiness to assist another in distress. Bekoff believes that a dog's sense of morality—altruism, tolerance, forgiveness, reciprocity, and fairness—is learned in play. He also suspects that over their fifteen-thousand-year history, dogs and humans used play to codify individual morality into a shared system of ethics. There's every reason to believe the process is ongoing. The dog looking at you with the oxytocin-gaze and asking that you lift the latch is also teaching you—or reminding you—to do unto others as you would have them do unto you.

Wood Thrush Songs, Herring Gull Drop-Catching, and Bowerbird Art: Play as the Roots of Culture

We . . . saw a raven perched near the edge of the canyon pick up a small rock with one foot and then by a combination of dragging and hopping, carry it to the edge where it used its bill to roll the rock over the edge. Then it looked down. The rock fell to a ledge twenty feet down. The raven immediately repeated the whole performance."[1] This account appeared in the article "Play in Common Ravens (*Corvus corax*)," by ethologists Bernd Heinrich and Rachel Smolker.

To an ornithologist—to just about anyone—the behavior invites questions. How might rolling small rocks over ledges to watch them fall assist an individual raven's survival? There is no known raven skill—such as flying or feeding—for which this might be training or practice. Ravens, we'd think, should have little to no interest in rocks to begin with. They don't eat them, they don't use them to build nests, and they have no obvious reason to push them over ledges. It likely wasn't courtship behavior, as no other ravens were present. Neither could it have been practice courtship behavior, since so far as ornithologists know, skill in rolling rocks does not enhance a raven's suitability as a mate. Because the raven was acting alone, it's hard to imagine how its behavior could have anything to do with social bonding. Heinrich and Smolker spend a few paragraphs in speculation, but in the end admit, "We see no obvious utilitarian purpose, proximate or ultimate, for this solitary behavior."[2] In that last sentence, bafflement shades into awe.

The Astonishing Avian Brain

Until recently, ethologists interested in play have paid little attention to birds. Since researchers from Groos onward have suspected that where there's play there's intelligence, and vice versa, it's a curious oversight, but explicable. It is owed in part to a misunderstanding of the avian brain that began with the work of German neurologist Ludwig Edinger.[3] A bird's brain has no neocortex. It has instead a "large pallial territory," structures that look like the tightly packed cloves of a garlic bulb, and which Edinger thought were basal ganglia. Since basal ganglia in mammal brains are responsible for motor control and coordination, and since the brains of birds seemed composed of nothing *but* basal ganglia, Edinger reasoned that birds were incapable of sophisticated behavior and could not, by most measures, be called intelligent.

In the twentieth century neuroscientists identified what Edinger regarded as basal ganglia as the *dorsal ventricular ridge*, or DVR, and a nucleus named the *wulst*. In 2020 a team of neuroscientists using a microscopy technique called three-dimensional polarized light imaging found that the DVR and wulst of both pigeons and barn owls functioned like the mammalian neocortex.[4] Other researchers found another similarity: neurons in the avian brain are like those in the brains of mammals, using similar pathways between regions of the brain and the same chemical neurotransmitters.

But there are also differences. Author and naturalist Diane Ackerman observes that birds have "a taut, quick vitality that seems almost too much for their tiny bodies to contain."[5] That vitality suggests that avian neuroanatomy might be economical, and many suspected that neurons in avian brains were densely packed. When in 2016 an international collaboration of researchers at universities in Prague, Vienna, Rio de Janeiro, and São Paulo managed to count them, they were surprised to discover just *how* densely packed. To their collective astonishment, they found that a parrot brain might have as many neurons as the brain of a midsize primate.[6] All of which is to say that a bird's brain is not, as Edinger believed, an undeveloped version of a mammalian brain. It is just as developed, but developed in a different way.[7] If by now you're thinking that the term *bird brain* should, by all that's right, be a compliment, many a neuroscientist would agree.

Avian Intelligence

Toward the end of the twentieth century, studies of avian behavior had shown that many bird species are remarkably intelligent by any measure. Some display astonishing feats of memory. A Clark's nutcracker will store pine seeds in thousands of caches spread over dozens of square miles and recall their precise locations months later, even when the seeds are covered in snow. Scrub jays, employing what is called episodic memory, can recall events that occurred at a specific time or place. Many birds show ingenuity. New Caledonian crows fashion probes and hooks from sticks, which they use to coax and pry grubs from their lairs. Other birds have impressive learning abilities. Pigeons can memorize more than seven hundred different visual patterns, categorize objects as "human-made" versus "natural," and rank patterns using what's called transitive inferential logic—deriving

117

a relation between items that have not been explicitly compared before. They can also discriminate cubistic and impressionistic styles of painting.[8] Animal psychologist Irene Pepperberg famously taught an African gray parrot named Alex to use over a hundred words and to identify shapes and colors. Alex may have been an unusually talented representative of a species known to possess impressive intellects. But other species of parrots also articulate queries and responses to humans in our languages, and in the right context—for instance, saying "good morning" only when meeting someone early in the day—and so persuading many animal researchers that they understand what "good morning" means.

Assuming that intelligence gives rise to play, we shouldn't be surprised that many bird species play quite regularly.

The Varieties of Avian Play

Johan Huizinga noted, "Woodcocks perform dances, crows hold flying matches, bowerbirds and others decorate their nests, songbirds chant their melodies."[9] Many naturalists have likewise been impressed by the sheer range of avian play. In 1977 Millicent Ficken, an American ornithologist with a special interest in birds' vocalizations and social behaviors, published the first survey of that range, an article titled "Avian Play." She included remarkable accounts of avian acrobatics. Corvids, she wrote, "frequently hang upside down on twigs or electric wires. They typically fall forward or backward from their normal perching position, hanging upside down by their feet with wings outspread, then often letting go with first one foot and then the other."[10] She cited accounts of ravens attempting and achieving feats that rival high-wire acts: "Balancing games were often conducted in connection with the upside down behavior and its many variations, the birds picking long thin branches, often balancing their way from a basal part to the end and then starting all over again." As if that weren't impressive enough, they "often they made this game more challenging by manipulating objects while balancing."

These behaviors may provide adaptive advantages. The bungee-jumping, dive-bombing corvids might be flight training, and their balancing on branches with or without objects might develop or maintain motor coordination. But Ficken's survey includes several accounts of bird behavior harder to explain—behavior with no evident adaptive advantage, either immediate or long term. She noted a report of an Anna's hummingbird riding down a stream from a hose; a group of common eiders gliding down a river's rapids and hurrying back to the spot from which they began to have another go; and Adélie penguins riding on small ice floes on a tide run.[11] It's hard to see how the hummingbird, eiders, and penguins can be doing much in the way of developing motor coordination, and their behavior seems to have no purpose related to survival or reproduction—or any purpose at all other than to experience the sheer kinetic pleasure of movement.*

---

* Darwin wrote, "How often do we see birds which fly easily, gliding and sailing through the air obviously for pleasure" (*Descent of Man*, pt. 2, 54). Watching a hawk circling in a cloudless sky, one is inclined to agree. Yet that inclination may be misplaced. If birds soared for no reason other than pleasure, we might expect that seed-eaters or fruit eaters would soar. But they don't. The only birds that soar are raptors, thus giving ornithologists reason to think they are hunting, or practicing hunting. Another species' behavior—the diving and swooping of female golden eagles—might also be mistaken for play. But ornithologists suspect it's a courtship display, a claim of territory, or both.

A Snowboarding Crow

More recent accounts of avian play include a much-viewed You-Tube video of a crow. Using the lid of a jar as a makeshift snowboard, the bird slides down a snow-covered roof, then flies back to the roof's peak carrying the lid in its feet and does it again. The headline for a piece in the *Atlantic* read, "Science Can Neither Explain nor Deny the Awesomeness of This Sledding Crow." Although science might neither explain nor deny its awesomeness, human inhabitants of northern latitudes might deny that its awesomeness is unusual. Heinrich and Smolker report that ravens in Alaska and northern Canada routinely "slide down steep snow-covered roofs, only to fly or walk back up and repeat the slide."[12]

The crow's use of the jar lid as a snowboard may qualify it for the second traditional play category: object play. Isolating avian object play from other behavior has proven challenging. When many birds push, prod, and otherwise manipulate objects, they seem to be both investigating and playing. Still, members of several avian species interact with objects in a manner that seems pure play. One such are herring gulls.

Researchers Jennifer Gamble and Daniel Cristol observed a flock of herring gulls that paused in their annual migration to frequent a mudflat in coastal Virginia at low tide. Herring gulls are known to drop clams to break them on rocks or hard surfaces to get to the meat inside—their own kind of foraging behavior. But Gamble and Cristol witnessed a rather striking nonforaging variation. A gull would carry a clam or other small object it its bill, toss it, catch it as it fell, then toss it and catch it again, flying all the while. The behavior was obviously neither foraging nor practice for foraging. Gamble and Cristol considered other explanations and, guided by a definition of play much like Burghardt's, eliminated all. The tossing and catching, they concluded, could only be play.[13]

Keas, the Bad Boy Birds of Play

Keas, the crow-size parrots that inhabit mountains on the South Island of New Zealand, are also particularly drawn to object play. In many animals, the impulse to play with objects is strong, but in keas it's *really* strong. One report describes a kea playing with a stick by chewing it, jumping on it, hitting it with his wings, and rolling under it as though it were an opponent in a play fight.[14] A controlled study provided a group of captive keas with a choice: food and a collection of inedible objects. Although the birds were hungry, they partook of the food only after spending several minutes playing with the objects.

Kea play is notoriously mischievous, as made evident to one hapless nineteenth-century European naturalist. This is Groos's telling: "He had with great difficulty collected a bundle of rare Alpine plants and laid them for a moment on a projecting rock. During his short absence a [kea] examined the collection and manifested his interest in botanical studies by pushing the whole bundle off the rock, never to be recovered."[15] It was a hint of things to come. The relation between human settlers and keas has long been troubled. In the nineteenth century keas attacked ranchers' sheep. Judy Diamond and Alan B. Bond, who conducted a four-year field study of the birds—and so know whereof they speak—call them "bold and curious, infuriatingly persistent, and ingeniously destructive."[16] With a strong curved beak they use as an all-purpose tool, keas are known to tear apart windshield wipers and trim from automobiles, and to take rain gutters and antennas from houses. By Burghardt's definition, the behavior is clearly play, albeit of a particularly destructive sort.

Kea social play is likewise a pernicious behavior, undertaken in groups and specific to a certain developmental stage. Diamond and Bond note, "Keas are fairly destructive at all ages," and with what seems a bit of puritanical finger wagging add, "The worst of this devastation is inflicted by mobs of idle juveniles."[17] Not an unreasonable characterization. The same authors report coming upon a particularly disturbing scene: "In one late night interaction a sizable group of immature birds formed a screaming circle around a pair of keas that were involved in a particularly vicious brawl, much like gang members gathered around a knife fight."[18]

Many bird species engage in social play that is rather more benign. When Australian magpies play fight, they try to peck their opponent on the side of the head. The opponent might attempt to escape, peck back, or stand bill to bill and create a stalemate. Sergio Pellis, here again borrowing a movement notation developed for choreography, found that play-fighting Australian magpies communicated their intentions with signals: a certain gait to initiate a play fight, a certain call to continue it, and a certain opening of its bill to end it.[19]

The Astonishing Acrobatics of Montagu's Harriers

The most spectacular avian play, as might be expected, is airborne. And the most spectacular of this may be that of the

Montagu's harrier, a medium-size raptor that in the words of one admirer has "a particularly graceful flight, with powerful and elegant wingbeats which give an impression of buoyancy and ease."[20] In 1996 ornithologist Massimo Pandolfi of the University of Urbino undertook a study of twenty-two nesting pairs in central Italy. Their play included pursuits, the birds taking turns chasing one another, sometimes in steep dives; play fighting, much like pursuits but more aggressive, with sudden bursts of acceleration and "talon presentation"; and the rather more serene communal soaring, several birds flying together in high circles.

All no doubt impressive, but not nearly so impressive as an adaptation of what is surely the most breathtaking avian behavior: *aerial food passes.* In a food pass, one harrier releases prey in flight so that a partner flying below can catch it midair. For Montagu's harriers, the behavior is typically part of court-ship, with males passing food to females, thus demonstrating that they are capable providers. Since the behavior has an immediate benefit, it's not play. Fledglings though, engage in a similar behavior that—since it does not serve courtship or use real food—*is* play. A young bird in flight will deliberately drop a small branch or tuft of grass so that another, flying a bit beneath and behind, might seize it midair.[21]

The Music of Birds

Many examples of avian behavior fit neatly in each of the three traditional categories of play. But play, being produced by that "sheer plenitude of vitality" is not constrained by categories, traditional or otherwise. Some bird play may, it seems, be found in their songs—which are not to be confused with calls. Birdcalls are brief utterances or cries of warning. Songs are longer and combine chirps, warbles, and whistles to produce a decidedly musical effect.

Birds learn songs as juveniles, in stages. First, the bird listens to the song of a parent or other bird. Making a first attempt, the juvenile produces what ornithologists call *subsong,* a series of sounds that have been compared to babbling in human infants. In four or five weeks, subsong develops into *plastic song.* Although its individual notes are recognizable as those of the mature song, they are intoned in no particular order. After two or three months (the length of time varying by species), the juvenile produces a song whose notes are both recognizable and arranged as they are in its species' mature song. This, finally, is *crystallized song.*

Birds sing both to attract mates and to establish territories, but sometimes they sing with no other songbirds in the vicinity—for no one, it seems, but themselves. Ornithologists call such song *undirected* or *solitary.* On some spring evenings just before twilight, I can hear a wood thrush intoning several low, flutelike notes followed by a complex trill. Sometimes he stops halfway through and begins again. It is not unlike, I imagine, a cellist practicing alone, working through a particularly difficult passage until he can perform it to his satisfaction. Upon getting the song right, the wood thrush may intone it yet again, not for practice (as none is needed) and not to impress (as no one else is present), but solely for its own enjoyment.

Avian behaviors that satisfy Burghardt's criteria for play also satisfy most criteria for culture. The Australian magpies' play fights compare to our athletic competitions, the Montagu's harriers' play-food passes to our dance, the wood thrush's song to our music.* Zoologist and psychologist Andrew Whiten called culture "all that is learned from others and is repeatedly transmitted in this way, forming traditions that may be inherited by successive generations."[22] British musician and composer Brian Eno defined it succinctly as "everything we don't have to do."

Culture is like play in four respects. First, it does not enable survival or reproduction in any immediate way; it has no obvious adaptive advantage. Second, it involves self-handicapping. In playing soccer we do not use hands; writing a sonnet we limit ourselves to fourteen lines; and when we compose for a cello, we compose for an instrument with four strings, not five. Third, much as self-handicapping in play compels an animal to adopt new moves, so self-handicapping in cultural practice invites or forces innovation. Fourth, culture is open-ended. We've mentioned that animals beginning a play session do so with no plan of when it will end; they cease playing only when they are injured, exhausted, or simply become interested in something else. So it is with culture. "No work of art is ever finished," says the aphorism. "It is only abandoned." It may be that the more open-ended a cultural practice is, the more play-like it is.

Consider the cultural practice we call the sciences. Roughly speaking, there are two sorts. Applied sciences are pragmatic, with an immediate application, while the basic or pure sciences are theoretical, with the goal of a better understanding of a subject. The latter sort may be most like play—and it has advantages.

---

* For a recent survey of animal culture, see Whiten, "Burgeoning Reach of Animal Culture."

Imagine two teams of humans trying to solve a problem, one in a pragmatic manner, the other in a playful manner. The pragmatic team works toward an end point. They know they've reached it when they have an answer to the problem, and they stop working. But the playful team, even after they've arrived at an answer, continue working simply because they are having fun. They are engaging in *tinkering*, which one dictionary defines as an "attempt to repair or improve something in a casual or desultory way, often to no useful effect." Sometimes, though, tinkering may produce an effect that is quite useful. The playful team may happen upon a second answer, one just as good as or better than the first. The principle underlying the playful team's engagement with the problem—that a given goal may be reached in many ways—is what Sir Patrick Bateson, emeritus professor of ethology at the University of Cambridge, called *equifinality*. It is, he said, a distinguishing feature of play, particularly the creative sort.

Isaac Newton was famously reported to have said that his own work, however significant to others, felt to him like the purest childhood amusement. In 1893 English-born natural philosopher W. Preyer made much the same comparison: "Man does not learn through any kind of instruction or study in later life anything like so much as the child learns in the first four years of his careless existence, through perceptions and ideas acquired in his play. . . . As I have previously spoken of the experimenting of little children as play, I may now mention the internal resemblance of their procedure to that of the naturalist."[23] The procedure of the naturalist may resemble that of the stone-dropping raven—or in some moods, our own. You approach a ledge. You look down. Having no pressing appointments, you pick up a small stone and toss it over. You watch it fall, bounce off an outcropping, and hear it hit bottom. Then you do it again. Perhaps the answer to why the raven was dropping the stones is the reason you and I might do it. What is *that* reason?

Groos opined that we find pleasure in simply *being a cause*. "How many of us want to scribble or whittle or do something with our hands all the time, to break a twig and chew it while we walk, to strike the snow off walls as we pass, to kick a pebble before us, to step on all the acorns on the pavement, to drum on the windowpane, to hit the wineglasses together, to roll up little balls of bread, etc."[24] The specificity of these details suggests the man knows what he writes about; it gives me an obscure pleasure to imagine Herr Professor Groos strolling the sidewalks of Basel on a cool autumn afternoon, occasionally interrupting his stride so he might step on an acorn for the sheer fun of it.

In late summer of 1878, an acquaintance of Darwin's told him of a pet monkey fond of looking at an object through an eyeglass and adjusting its focus by moving it to closer and farther from that object. Darwin recalled that he had been unable to teach his two-year-old grandson the same technique and surmised, "A child just under two years is inferior in intellect to a monkey."[25] By this time Darwin was late in his career, and he had left work on animal intelligence to others, particularly his friend and protégé George Romanes, to whom he'd given his notes on the subject and with whom he maintained a lively correspondence. In one letter, Darwin asks, "Have you ever thought of keeping a young monkey, so as to observe its mind?" Romanes replied that he had. In a series of exchanges Darwin advised the younger man on how to acquire a suitable primate, and both made half-serious suggestions for a comparison study in which Romanes might raise the animal alongside his daughter, then not a year old. The exchange concluded with a letter of December 17, 1880, in which Romanes announced triumphantly, "I have now got a monkey. Sclater let me choose one from the Zoo, and it is a very intelligent, affectionate little animal." Then, in the sentence following, he acknowledged a minor difficulty: "I wanted to keep it in the nursery for purposes of comparison, but the proposal met with so much opposition that I had to give way."[26]

One assumes that the opposition arose from Mrs. Romanes. Fortunately, for both the general advancement of scientific inquiry as well as the Romanes marriage, Romanes's sister Charlotte, who lived near him, was willing to keep the monkey and take notes for her brother, giving particular attention to behavior that signified intelligence. On December 18, 1880, a tufted capuchin (*Cebus apella fatullus*) was delivered to her well-appointed apartments in the City of Westminster, where it resided for ten weeks. During that time the monkey enjoyed much of what Groos would call the pleasure of having an effect, and quite often that effect was chaos.

Charlotte Romanes's entry of December 21:

I notice that the love of mischief is very strong in him. Today he got hold of a wine glass and an egg-cup. The glass he dashed on the floor with all his might, and of course broke it. Finding, however, that the egg-cup would not break for being thrown down, he looked around for some hard substance against which to dash it. The post of the brass bedstead appearing to be suitable for the purpose, he raised the egg-cup high over his head and gave it several hard blows. When it was completely smashed he was quite satisfied.[27]

The craving for fun or sensation is the immediate reason we toss stones over ledges and crush acorns underfoot, and the reason Ms. Romanes's houseguest broke the wineglass and the eggcup. But what is the ultimate reason? What is the adaptive advantage of such activities? We might suppose that at least one is education. The raven dropping stones over a ledge and watching them fall might be learning about gravity, aerodynamics, and the speed of falling objects. The tufted capuchin may have gained knowledge of the forces of kinematics, the weights of objects, and the strengths of materials.

In an iconic scene in Stanley Kubrick's *2001: A Space Odyssey*, a man-ape sits idly among a field of tapir bones. He has no evident purpose; he is only mildly interested in seeing what happens when bone strikes bone. He is playing. But then he discovers that when he brings a thigh bone down with enough force, it can break and shatter other bones. He has a sudden epiphany: the bone may be used as a weapon.

Culture Is Like Play Because It Begins in Play

The scene from *2001* suggests that play gives birth to human tool use and perhaps engineering. Huizinga was convinced of much the same, declaring play's influence on much culture profound and seminal: "Ritual grew up in sacred play; poetry was born in play and nourished on play; music and dancing were pure play."[28] Given the absence of a historical record for such origins, these hypotheses may be difficult to support. But in his book *Wonderland: How Play Made the Modern World*, Steven Johnson makes a persuasive case that major developments and advances in civilization were driven not by necessity, but by the desire for novelty and new ways to play.

As to an animal culture developing from play, there is one well-documented case.[29] In 1988 a female bottlenose dolphin named Billie was rescued from a polluted creek near Adelaide, Australia, and housed temporarily at a local marine mammal park. There she kept company with five other dolphins. The others, so they might make public performances, had been trained to "tail-walk"—forcing most of their bodies vertically out of the water and maintaining the position by vigorously pumping their tails. Billie was never given such training, but she observed the others as they rehearsed and performed. Some weeks after her rescue she was returned to the wild. She was later sighted in the open ocean, tail-walking with a female dolphin named Wave. Other dolphins in the pod were sighted tail-walking, too. Since the behavior was nonfunctional, voluntary, characterized by repeated but varied movements and its performers were well-fed, safe, and healthy, it met Burghardt's definition for play. Since it was learned from others, was repeatedly transmitted and formed traditions, it also met Whiten's definition for culture.

The researchers who witnessed and documented Billie's and Wave's behavior stated, "The first wild tail walk was observed in 1995."[30] But one observation was made a century earlier, recounted by none other than Groos:

> Every seaman is delighted to see a school of dolphins. The cheery travellers hurry along through the swelling waves in a long and regular train, pursuing their way with a speed that suggests a race, and with leaps of wonderful agility. Their glittering bodies rise in the air in fine curves from one to two yards wide,

fall headlong into the water, and soon spring up again, carrying on the game. The jolliest of the flock turns somersaults in the air, turning up their tails in a most comical manner. Others fall flat on side or back, and still others remain bolt upright, dancing along with the help of their tails until they have made three or four forward movements.[31]

There are several lessons here. One is that since dolphins have learned to tail-walk without humans training them, again we may overvalue our talents and undervalue animals'. Another is that credible accounts of animal behavior are too often overlooked and lost. Still another is a reminder that there is much we don't see. Animals are regularly engaging in behaviors unobserved by scientists; some of it, such as dolphin tail-walking, is likely to be remarkable.

The Culinary Skills of Field Mice

Like many New Englanders, I heat with a woodstove, on cold winter nights finding it if not utterly necessary, then at least particularly welcome. Returning home one evening, I detected an aroma, particularly noticeable near the stove, still warm from the previous night's fire. Not unpleasant, it was redolent of buttered popcorn. Curious and more than a bit mystified, I removed the cover to see, in a crevice near the iron heating plate, a cache of ten or twelve pine nuts, each roasted to a lovely golden hue.

I had suspected mice were somewhere in the house—they are not uncommon in old New England farmhouses—so I wasn't surprised to find that a bag of pine nuts in a kitchen cabinet had been gnawed open. A mouse had apparently found the nuts and carried them, one or a few at a time, through several rooms to the woodstove, where it had discovered—or rather invented—an efficient means to prepare supper for a winter night. This remarkable achievement showed a discerning preference for cooked food over raw. (Not that the mouse knew this, but cooking increases the energy available from raw food.) It also required imagination, foresight, planning, and the not-inconsiderable self-discipline necessary to carry raw nuts in one's cheek while resisting the temptation to eat them. Finally, this achievement exhibited a bit of Yankee parsimony. The use of fire has long been regarded as a mark of civilization and a point of distinction between animals and humans. But the mouse didn't need to go through all the trouble of harnessing fire. It had harnessed me.[*]

I had forgotten about the pine nuts until I happened across a 2015 study by researchers Felix Warneken and Alexandra G. Rosati, conducted among chimpanzees from the Tchimpounga Chimpanzee Sanctuary in the Republic of the Congo. The chimpanzees were supplied with two Tupperware-like containers in which they could store raw sweet potato slices. With the chimpanzees unawares the researchers refilled the raw potato slices in the first container with more raw potato slices, but exchanged the raw slices in the second container with cooked slices. Soon, the chimps learned to store their raw potato slices in the second container, having lost all interest in the first. The chimps'

---

[*] I can't be certain that the mouse chose the stove as a caching place because it warmed and roasted the nuts, and I conducted no controlled experiment. But the house offers innumerable locales for storage, and I found them nowhere else. I removed the cache twice; it was twice replenished, the nuts never left long enough to burn. Many animals, given a choice, prefer cooked food to raw, and mice are known to plan. Sweis et al., "Mice Learn to Avoid Regret."

behavioral adjustment led Warenken and Rosati to conclude that the mental capacities necessary to cook, "causal reasoning, self-control and anticipatory planning," are not unique to humans.[32] Others contended not that the chimpanzees lacked mental capacities to cook (upon which they were agnostic), but that the experiment had not shown them. It had shown only that chimpanzees could learn to trade raw sweet potato slices for cooked sweet potato slices.[33]

The mouse that made use of my woodstove seemed to have an edge on the chimps. It, too, demonstrated "causal reasoning, self-control and anticipatory planning," but whereas the chimps had been provided materials necessary and steered toward a behavior that mimicked cooking, the mouse, with neither guidance nor prompting, had recognized that the heat from the woodstove offered a means to make the nuts more palatable. It had discovered or invented a means to roast the nuts and, it would seem, fully understood that it was not trading raw for cooked nuts, but rather that the nuts it placed in the woodstove were the same nuts it would retrieve from it later. The mouse was, by most definitions, actually cooking.

As an expression of creative skill and imagination, cooking is an art. But since it requires knowledge of thermodynamics and chemistry, it's also a science. If more than one mouse was caching the nuts, and if it (or they) had learned the behavior from the first mouse to cache, then all were engaging in a behavior that meets definitions of culture.

Perhaps wild mice, when not enjoying roasted pine nuts, are developing their skills well out of our sight in murid culinary schools. I think it more likely, though, that animals are all the time behaving in astonishing ways that we simply fail to notice, and that no one, at least no one with the right resources, has thought worthy of study.

Some anthropologists have argued that for a behavior to be defined as a culture, it must be more than inheritance of behavioral traditions acquired through social learning. Those traditions, they say, must show improvements or refinements over several generations. Even by this demanding standard, many animals may be said to possess cultures. In recent years, members of the primate archaeology project led by Michael Haslam at the University of Oxford discovered stone tools at sites in West Africa, Brazil, and Thailand. The implements are crude, and you and I might mistake them for ordinary broken stones. But they are remarkable—*astonishing* is not too strong a word—in that they were not made by humans or by human ancestors. Rather, they were made and wielded by chimpanzees, capuchins, and macaques. They are tangible evidence that nonhuman primates have developed cultures with a stone-based technology. Those animals have, in other words, entered their own Stone Age.[34]

Haslam notes that it's possible chimpanzees, macaques, and capuchins haven't reached the limit of their technological capabilities, but since "smaller populations cannot spread and sustain complex technologies as well as larger groups,"[35] by shrinking their numbers through hunting and habitat destruction, we humans may have limited it *for* them.

Assuming that chimpanzees survive the myriad threats to their survival, in half a million or a million years, might their termite-fishing tools become instruments as refined as those of any human machinist? Might whale songs, in the fullness of evolutionary time, become epic poems greater in their way than any written by humans or symphonies against which Bach's Mass in B Minor would sound like "Chopsticks"? And what of bower-birds—whose constructions are so carefully fabricated that the first European naturalists to happen upon them assumed they had been made by aboriginal children? Given half a million or so years, could bowerbirds evolve skills that would enable them to design and construct the avian equivalent of cathedrals?

In a time when the litany of environmental woes seems to grow ever longer, the worry that human destruction of animal habitats may inhibit the development of animal culture might seem altogether too hypothetical, too abstract, and too distant to merit our concern. Perhaps it is. But at least it prompts us to consider exactly what we *should* worry about, and exactly where to position what ethicists call our "moral horizon."*

And as for animal play? The possibility that it might shape some animal cultures and give birth to others is another reason it's worth our attention.

Natural Selection, Play, and Beauty

Darwin noted that a male peacock's extravagant tail feathers make it difficult or impossible for the bird to escape predators. He allowed that natural selection could not explain the feathers' evolution, and accounted for them with what is sometimes called Darwin's *other* theory—his theory of *sexual selection*, the

---

* Although we have not made sustained or concerted efforts to preserve the culture of birds, one bird for a brief period may have preserved part of our own. Darwin: "The partial and complete extinction of many races and sub-races of man are historically known events. Humboldt saw in South America a parrot which was the sole living creature that could speak the language of a lost tribe." *Descent of Man*, pt. 1, 236.

selection produced from the predisposition of one sex in a species for certain characteristics in the other sex of the species. Sexual selection, so Darwin posited, is responsible for much of the natural world's beauty.

But Darwin's natural selection also produces beauty. When an insect lands on an orchid and sips its nectar, the orchid's pollen brushes off onto the insect. The insect carries the pollen to another orchid and brushes the pollen onto that orchid's stigma, and eventually the orchid will produce hundreds of thousands of seeds. What attracts the insect to orchids in the first place is the shape, symmetry, and color of their petals. The orchid's beauty, the adaptive advantage that enables it to reproduce, is a product of natural selection.

Thus to the similarities between natural selection and play we may add another. The play of birds—some of which may be culture or the beginning of culture—also produces beauty. The solitary play of the wood thrush is harmonious. The object play of drop-catching herring gulls is elegant. The social play fighting of Australian magpies is symmetrical and balanced. And the exquisitely coordinated aerial food passing of Montagu's harriers has an athletic grace that is, for anyone fortunate enough to witness it, astonishing.

CHAPTER 7

Memes and Dreams: Dreaming as Playing without a Body

A cuttlefish's camouflage renders it indistinguishable from its surroundings and background. The change is rapid, a flickering across the skin in seconds, and the effect is otherworldly, hallucinatory. The cuttlefish achieves it by dilating and contracting chromatophores, tiny elastic sacks of pigment in its skin cells. How much of this activity is deliberate and controlled is unclear, but it seems to work even when the cuttlefish is unconscious. Peter Godfrey-Smith, a full-time philosopher of science and a part-time scuba diver, was swimming in waters off Australia when he came upon a sleeping cuttlefish. It was nearly still, but the hues of its skin were ever shifting.

> I wondered if this was a cuttlefish dream—I was reminded of dogs dreaming, their paws moving while they make tiny yip-like sounds. He made almost no movement, except small adjustments of siphon and fin that kept him hovering in the same place. He seemed to be maintaining as little physical activity as possible, except for the ceaseless turnover of colors and patterns on his skin.[1]

Despite the work of ethologists, neuroscientists, and studies on hundreds of species, much about play in animals remains a mystery. But it's not the only one. Ethologists have long been intrigued and often bewildered by another behavior—one that Godfrey-Smith may have observed in that cuttlefish, and which is also widespread among animals: dreaming. As we'll see, play and dreaming are much alike, so alike in that it's stretching matters only slightly to say that one could substitute the word *dreaming* for *play* for long passages in this book, and they'd be just as valid.

Dreaming is difficult to define because like play, it's an experience with hazy boundaries, occupying one end of a behavioral spectrum with waking thought at the other, and innumerable, subtle gradations of mind wandering and daydreaming between them. It's perhaps not surprising that, like play, dreaming has been defined many different ways; one survey counted twenty definitions appearing in the same journal over nine years.[2] Dreaming is difficult to study because like play, it's a subjective experience, not easily observed and resistant to methodological rigor, precision, and objectivity.

The Play Theory of Dreaming

Once we acknowledge the difficulties of defining and studying dreams and set them aside, we're left with dreaming itself, and we find that it also has much in common with play. The similarities have long been noted by psychologists and cognitive scientists. But one researcher has gone further, drawing those similarities into rather strict alignment. Kelly Bulkeley is a psychologist of religion and director of the Sleep and Dream Database, a digital archive and search engine for the scientific study of dreams. Bulkeley thinks it is no happenstance that dreaming looks like play—he believes it *is* play. More precisely, it's an imaginative sort of play enacted within the subject's mind. The idea he calls the *play theory of dreaming* draws upon human psychology and evolutionary biology. Especially relevant to our interests is that it makes use of the ideas of Gordon Burghardt. With only slight adjustments, says Bulkeley, Burghardt's definition of play might as easily describe dreaming.

Burghardt's Idea of Play Applied to Dreaming

Burghardt called play nonfunctional. Since dreams do not directly cause any waking behavior, they can't directly fulfill any function related to survival or reproduction. In fact they can't in any straightforward or immediate way affect anything at all outside the mind of the dreamer. Dreaming then, like play, is nonfunctional.

Burghardt called play voluntary. The word *voluntary* may imply consciousness, and we may think of any voluntary behavior as a conscious choice. A dreaming animal is unconscious and does not consciously choose any particular experience within its dreams. Several neuroscientists have proposed that dreams—and therefore the experiences they create—are nothing more than the product of the random firings of synapses.[3] This idea is reminiscent of Herbert Spencer's surplus energy hypothesis of play. But even if those firings are random, they are nonetheless an autonomous behavior that arises from within the animal. Perhaps not a choice, but it is at least *like* a choice. On the other hand, if consciousness is not a prerequisite for choosing, and an unconscious mind *can* choose, then dreaming might be said to be voluntary, with the word *voluntary* here meant in its usual sense.

Burghardt said play was characterized by repeated but varied movements. If those movements are movements of the body, the criterion may not apply to dreaming. But if the movements are those experienced by the dream self within the dream, we may find their equivalent. Situations within a recurring dream often repeat, but almost always with variations. Consider the proverbial actor's nightmare, of being caught onstage without knowing one's lines. The nightmare recurs, but details change in each iteration—different lines, different performances. Or consider the conscientious student's nightmare—being utterly unprepared and late for a final exam. That nightmare, too, recurs, but again, with each iteration details are altered—different final exams in different classrooms. In both cases, the recurring elements call upon the dream self to repeat its responses, but the details unique to each iteration call upon the dream self to vary them. Dreaming then, like play, is characterized by movements that are repeated but varied.

Burghardt said play occurs only when the subject is well-fed, safe, and healthy. Since an animal that is ill, fearful, or hungry will experience disrupted sleep, its dreaming will also be disrupted. Dreaming then—at least undisturbed dreaming—occurs only when the subject is healthy, safe, and well-fed.

To Burghardt's definition of play—behavior that is nonfunctional, voluntary, characterized by repeated but varied movements, and occurring only when the animal is healthy, safe, and well-fed—we may now add a corollary. If those repeated but varied movements occur only in the animal's mind, the definition also describes dreaming.

"I Strongly Suspect That Ferrets Dream"

Since scientists cannot see animal dreams, they cannot study them directly. They can, however, study animal dreaming by observing a dream's outward manifestation: a *dream-enacting behavior*. A sleeping dog may move its legs as if it were running and produce a hoarse, half-suppressed bark. In *The Passions of Animals*, Edward Thompson observed, "When the impressions of the dream assume a particularly vigorous and distinct character, they affect the slumbering voice and limbs, and thus prove most satisfactorily and clearly that animals really do dream."[4] He alluded to accounts of dreaming behavior in horses, elephants, and several species of birds—among them storks, canaries, and eagles. Darwin, drawing on his own observations and those of others, concluded, "Probably all the higher animals . . . have vivid dreams, and this is shewn by their movements and voice."[5] And Romanes declared most memorably, "I strongly suspect that ferrets dream."[6] But exactly what are dogs, horses, canaries, and ferrets dreaming about? To learn what humans dream, researchers have traditionally relied on a report from the dreamer upon waking, a practice not feasible with subjects unable to provide such a report. But there may be other ways.

Sleep-Talking Birds

Human talk in sleep is termed *somniloquy*. Its brief utterances and fragments of sentences are often too incoherent to serve as a report on any dream, but may nonetheless give hints as to its nature. Somniloquy is a remarkable behavior, and one we may add to the ever-lengthening list of those not unique to humans. The more loquacious species of birds also talk in their sleep. How, one might ask, does one know a bird is sleeping? Although it stays upright, keeping mostly still, a sleeping bird bobs its head and makes slight movements of its wings. The behavior is well-known to people with avian pets, with many YouTube videos showing them sleep-talking. (Naturalists have long reported on avian somniloquy. Romanes wrote of sleep-talking parrots.) As somniloquy now offers insights into the dreams of humans, so it might one day offer insights into the dreams of birds.[7]

Several studies of dreams have used *electroencephalography*, the measurement and recording of electrical activity—what are termed neural firing patterns—in different parts of the brains of human subjects. One such study found that a subject's neural firing patterns during sleep—thought to be manifestations of dreaming—are often identical to neural firing patterns that occurred during that subject's waking experiences shortly before. The suggestion is that some dreams reactivate or replay recent memories. If this is what's happening, then the phenomenon suggests intriguing possibilities for research. By comparing such patterns in a chimpanzee when it's awake and again when it's sleeping, for example, neuroscientists might learn not only that the chimpanzee is dreaming, but also what it's dreaming about.

The difficulty with some such studies is that they are inherently inhumane, requiring that a transmitter be implanted in the brain. Any system that records the electroencephalogram of freely moving animals requires lesions. But for the class of mollusks that includes octopuses, squids, and cuttlefish, studies of dreaming might possibly be undertaken with no such interventions— because their neural activity is visible. As a 2001 paper noted, "The body patterns demonstrated by octopuses may be representative of patterns correlated with waking activities."[8] The body pattern of an awake octopus when it is mating is likely different from its body pattern while foraging, but may be quite like the body pattern of that octopus when it is dreaming of mating. Octopuses (and cuttlefish) might be read like a book—if we only understood the language in which it was written.

If dreaming is a kind of play as Bulkeley suggests, then it follows that dreaming and play may have the same adaptive advantages. Before we discuss that possibility, we'd be prudent to consider the nature of the behavior that makes dreaming possible.

The Mystery of Sleep

Since animals dream only when asleep, many neuroscientists suspect that sleep may be necessary to dreaming. Evolutionary biologists suspect that sleep may have evolved first, with dreaming a behavioral offshoot. They have not agreed on the adaptive advantages of sleep, but they suspect it has them for the same reasons they think play does. One is that since the behavior is common and widespread across taxa, natural selection seems to regard it as a sound investment. The other is its rather striking *dis*advantages. For many animals, and certainly for most mammals, in sleep the eyes are closed, muscles are relaxed, the nervous system is relatively inactive, and consciousness is suspended. A sleeping animal is incapable of undertaking activities necessary to its survival. It's also vulnerable to predators.

The Adaptive Advantages of Sleep

Evolutionary biologists reason that natural selection would not have selected for sleep in so many species unless its disadvantages were countered by equally significant advantages. A great many have been proposed, including maintaining a stable brain temperature, processing memories, and promoting brain development. Animals' sleep is astonishingly various, differing, for instance, in both its duration (horses sleep a mere three hours per day, the northern night monkey upward of seventeen) and the role of parts of the brain (female frigate birds can sleep on the wing, with one hemisphere asleep and the other awake). As Spinka and company note, "No single hypothesis fully explains the diversity in which sleep manifests across different taxonomic groups."[9] They have a hypothesis that partially explains that diversity, if not in all animals, then at least in mammals and birds.

Most mammals and birds cycle through two types of sleep. One is rapid eye movement (REM). It's characterized by irregular respiratory and heart rates, occasional muscle twitching, reduced muscle tone, increased brain temperature, and, per its name, rapid eye movements under closed eyelids. The other is non-rapid eye movement (NREM), which is characterized by lowered brain temperature, and respiratory and heart rates that are slower and more regular than they are during wakefulness. Spinka and company suggest that when the two sleep states first appeared, they bestowed a particular adaptive advantage—perhaps maintaining a high, stable brain temperature, processing memory, or promoting brain development. Whatever the advantage, Spinka and company believe that it must have been significant because sleep has endured. It has also diversified. The range of types of sleep exhibited by animals in our time—horses, northern night monkeys, and frigate birds among them—are evidence that since the appearance of its initial adaptive advantage, sleep has evolved a great many others.

Mammals and birds are not the only species that slumber. Most animals experience a state that neurobiologists agree is sleep. Reptiles, fish, and even insects demonstrate what are called sleep signatures, that is, "quiescence, a specific sleep posture [and] reduced responsiveness,"[10] and electroencephalograms on these show lowered brain activity. Although only mammals and birds seem to experience REM and NREM sleep, neuroscientists have identified two sleep states—and possible correlates of REM and NREM—in some reptiles and some fish. Yet sleep is not confined to vertebrates; fruit flies, honeybees, crayfish, and—most remarkably—jellyfish seem to sleep. This last is rather a surprise since sleep is controlled by a central nervous system, and jellyfish have none. They do, however, have neurons, and jellyfish may sleep because their neurons cycle through periods of low activity. If the land of Nod has an outer perimeter, it's difficult to be sure where to draw it. Even plants have circadian rhythms.

The wide range of forms of sleep has evolved since its initial appearance suggest an equally wide range of adaptive advantages. One of them—in animals with sufficiently complex brains—might be dreaming.[11] Dreaming, as Bulkeley's definition notes, does not in any obvious way enable an animal to survive or reproduce. Like play and sleep, it is patently disadvantageous: an imprudent use of time and energy at best, and downright dangerous at worst. Yet since many animals dream, evolutionary biologists assume that it must have adaptive advantages. There are many ideas as to what they might be. Since human dreaming is easiest to study, it's been the focus of most.

The Adaptive Advantage of Dreams

One of the most widely accepted theories at present is the *threat simulation theory*, put forth in 2000 by Finnish researcher Antti Revonsuo.[12] It bears resemblance to Groos's practice hypothesis of play and its various iterations, positing that the adaptive advantage of dreams that contain frightening elements—such as dreams in which the dream self is chased or attacked—are simulations of possible threats in waking life and provide exercise for the cognitive mechanisms necessary to perceive and avoid them.

Another theory of dreaming is the *social simulation theory*, which explains some dreams as an "immersive spatiotemporal simulation." Or, in the words of one paper, "a simulation of human social reality, simulating the social skills, bonds, interactions, and networks that we engage in during our waking lives."[13] This, too, may sound familiar, since it resembles the social bonding hypothesis of play and its variants—that play enhances social competence, social negotiation, social assessment, and social manipulation.

Dreams as Training for the Unexpected

Since Bulkeley's play theory of dreaming posits adaptive advantages for both dreams involving threats and dreams involving social situations, it's more comprehensive than the threat simulation theory and the social simulation theory. Spinka and company's training-for-the-unexpected hypothesis—a theory of the adaptive advantages of play—may explain still more. Recall that

Spinka and company posit that animals welcome the unexpected, and that some, finding no unexpected in the offing, may create their own. We humans create our own unexpected in dreams. In a dream, we might open a door and be surprised by what's behind it. That door and the thing behind it were created in the dream by us—or, more specifically, by our unconscious. In such a dream we are both producing the surprise and being surprised by it.

Recall, too, that Spinka and colleagues proposed that the most essential feature of play may be self-handicapping. It certainly figures in many dreams. The dream of the actor or student whose dream self is unprepared doesn't end with that realization. Quite often that's only where the dream begins. It may continue with the dream self obliged to go onstage or begin the exam.

If dreams do serve as a rehearsal for some eventuality in our waking lives, we might wonder why so many dreams seem to have so little to do with those lives. Of what practical use is a childhood memory of a place we'll never visit again or a meeting with a person long dead? What might we make of some other dreams that seem to have nothing whatsoever to do with our waking existence, past or present? Dreams that are so bizarre, so utterly and categorically surreal, that there's no conceivable eventuality for which they might serve as practice?

Dreams, Play, and Creativity

Perhaps some dreams are not rehearsals. Perhaps they are the products of a mind released from the constraints of reality and so able to freely explore and imagine. In dreams, says neuroscientist

Matthew Walker, "No longer are we constrained to see the most typical and plainly obvious connections between memory units," and thus liberated "the brain becomes actively biased toward seeking out the most distant, nonobvious links between sets of information."[14] In dreams we are attracted to abstractions, to novelty and hyper-associativeness. We are, one might say, more playful.

That playfulness is not lost upon waking. Research subjects roused from REM sleep (and so perhaps from a dream) solve word puzzles faster than they do when fully awake, and evidently with less deliberation and more intuition. Recall that it was upon waking one morning that Jaak Panksepp was struck with a suspicion that rat chirping was laughter. (Incidentally, it was Panksepp who suggested that both play and dreaming behaviors might use the same neural pathways.)[15] One wonders whether the thought arose in a dream. If it did, it might be added to innumerable examples of unconscious inspiration. Paul McCartney came upon the melody for "Yesterday" in a dream.[16]

As for dreaming and creativity in animals? Evidence suggests that one species finds inspiration much as the bassist from Liverpool did. Zebra finch sing when they are sleeping. One study discovered that while they sleep, their brains spontaneously reproduce the activation patterns they make when they sing during the day—a finding that strongly suggests they are dreaming about singing. More astonishing is that their subsequent waking songs are nearer mature song—which is to say, improved. It seems that zebra finches not only sing in their sleep. They rehearse.[17]

*　　*　　*

Play offers a means to escape and transcend not only social and cultural constraints. In the mind of the player, it's a means to escape and transcend physical constraints—the body, even space and time. In play, that escape may be qualified; in dreams, it is complete.

I'm probably correct in suspecting that most fans of pro wrestling are aware that any given match is not real fighting, but a performance. When they cheer and jeer, they participate in that performance, making for communal pretending. The wrestlers and fans are aware of two states of affairs—one real, another illusory—and as the match proceeds, they direct that awareness from the real to the illusory and back again. Groos notes that many animals are capable of the same behavior and indulge in it quite readily. In *The Play of Animals* he writes, "If, then, in conscious make-believe, in the young dog, for example, that begs his mistress to reach out her foot and then falls upon it with every sign of rage, but never really biting it, the connection between the pretended I and the real I underlying it is preserved in spite of the division of consciousness."[18]

Owing to the dual awareness of a playing animal, like that of the dog who feigns biting his mistress's foot, the player always remains aware of the world outside its play. Like a bird tied to a string, it can only fly so far. But dreaming may be different. It's true that a dual consciousness exists in waking dreams, that is, dreams in which dreamers know they are dreaming. It's also true that the waking world may intrude upon and affect dreams, as when a noise is woven into a dream without interrupting it. But in most dreaming the world outside is left behind, that string is cut, the bird is released.

In Bulkeley's view, dreaming is play enacted within the mind and freed of the body, and so freed of the body's needs and limitations. A compelling view to be sure. It may be instructive to turn it upside down, or—taking a cue from the self-handicapping chimp a few chapters back—turn ourselves upside down, thereby experiencing a momentary disorientation but gaining the advantage of a new perspective, and look again. If dreaming is disembodied play, then perhaps play is embodied dreaming. If in dreaming we are playing without our bodies, then perhaps in playing we are using our bodies to dream.

Play operates in dreams, and natural selection operates in a place just as intangible—the realm of language. Darwin noted that words and grammatical forms struggle for life, that languages may show variation, blend and cross-fertilize, spread widely or diminish and suffer extinction.[19] The idea is widespread, but it was little examined until 2008, when evolutionary theorists found that new languages evolve much like organisms—not by slow changes, but in sudden spurts.[20] And so we may posit another similarity between play and natural selection. Both can operate without a material form.

In the final chapter of *Origin of Species*, Darwin observed that we would do well to "regard every production of nature as one which has a history."[21] Animal play is a production of nature and has a history or, as we'll see, several histories. By tracing those histories, we may gain a larger, comprehensive, and more holistic understanding of play. So far we've attended to three of Tinbergen's questions as applied to play: its possible adaptive advantage or advantages, its development in an individual animal, and the features necessary to perform it. It's time we attended to the fourth. How did play evolve?

CHAPTER 8

The Evolution of Play

Charles Darwin did not preserve drafts of *On the Origin of Species*, and several of the few pages we have are those whose blank sides were used by his children for drawing paper. To Darwin, the usefulness of these sheets was short term, and once he had written out a fair copy of the manuscript, they became a surfeit. They were, though, capable of being repurposed and were repurposed by his children for play. Darwin's thrift mirrored that of the theory for which he is renowned. Natural selection is frugal, too. Confronted with an evolutionary pressure, it seldom produces a new design from scratch, but rather refashions an existing design. Darwin described that process, using the bones in an animal's appendage as an example: "[They] might be shortened and widened to any extent, and become gradually enveloped in thick membrane, so as to serve as a fin; or a webbed foot might have all its bones, or certain bones, lengthened to any extent, and the membrane connecting them increased to any extent, so as to serve as a wing." Yet the fundamental pattern, he noted, is unchanged: "In all this great amount of modification there will be no tendency to alter the framework of the bones or the relative connexion of the several parts."[1]

Recall that Burghardt's surplus resource theory proposed that play develops in the life of an individual animal in a way that is also economical. Primary process play that appears in a young animal as "excess metabolic energy" is repurposed as secondary process play that enhances its physiology, then repurposed again as the tertiary process play that supplies practice for the innovative behaviors that bring adaptive advantages. Burghardt proposed that this "waste-not-want-not" development in an individual animal is also evident in the evolution of the species to which that animal belongs. Burghardt's theory posits inflection points in that evolution—a point at which primary process play became secondary process play, and another at which secondary process play became tertiary process play.

Imagine a newborn capuchin monkey—not a present-day capuchin, but its distant ancestor. Its squirming and wriggling were primary process play, nonfunctional, and with no benefit to the animal. But many generations later, that capuchin's descendants began to channel and direct that activity and develop play fighting. It had become secondary process play, and it benefited the capuchins by maintaining motor coordination. Over still more generations, the descendants of those capuchins gradually refined their play fighting to such a degree that it could be called tertiary process play. It benefits present-day capuchins, among other ways, by supplying practice in assessing their opponents.

As applied to natural selection, the surplus resource theory is open-ended. It does not posit that play must be practice for hunting or foraging. Nor does it suggest that play must be practice for sex, self-medication, acquiring general social competence, defusing conflict, or training for the unexpected. It does, however, allow that it *might* yield any or all those adaptive advantages, as well as others as yet undefined, which in both kind and number may be limitless.

Like Burghardt, the Pellises suspected that the development of a behavior in an individual animal retraces the evolutionary history of that behavior in the species itself. The development of play in individual rats, they thought, was a palimpsest of the evolution of play in the species. It made for an interesting and reasonable hypothesis, but the Pellises knew it was limited in scope, derived after all, from the studies of only one animal. They wanted to understand the evolution of animal play more generally. To do that, they would need to step back and take in a much-wider view.

When evolutionary biologists wish to trace the development of a certain feature over time, behavioral or physical, they begin by looking at that feature in one species, then the same feature in a related species. Next, they hypothesize the presence of that feature in both species' presumed ancestors, working back along the phylogenetic tree, stem to branch to larger branch, to the perhaps hypothetical but nonetheless probable ancestor of both.

Until the mid-twentieth century, evolutionary biologists presumed that the more alike any two species looked, the more closely they were related and the more recently their evolutionary lines had diverged. Lately though, scientists' achievements in whole-genome sequencing have shown that such appearances can be deceiving. Chimpanzees look more like orangutans than they look like humans, yet they are more closely related to us. Researchers have discovered that while orangutans and humans share about 97 percent of their DNA, chimpanzees and humans share approximately 99 percent. At the level of phylogenic classes, there are more surprises. Fungi look more like plants than animals, yet are nearer relations to the latter. The mushrooms in your salad are more closely related to you than to the parsley that garnishes them.

The Promise of Cladistics

Resemblance of a general sort, then, is not a reliable indicator of recent common ancestry. Evolutionary biologists consider shared, measurable characteristics (even if not easily seen) to be stronger indicators, and maintain that the greater the similarity in these characteristics compared to the overall differences between two species, the more recently they diverged from a common ancestor. This assumption underlies the means of determining ancestry called *cladistics*, the word derived from the Greek *klados*, or "branch." In cladistics, the lineal descendants of a common ancestor, as well as that ancestor, compose a *clade* (rhymes with *blade*), and a visual representation of a particular clade or set of clades—this is contemporary evolutionary biology's answer to the phylogenetic tree—is a *cladogram*.

The Pellises wanted to create a cladogram that would show the evolution of play fighting in rats and their near relations, all murids—a family that counts among its members Mongolian gerbils, prairie voles, and house mice. They began by measuring play fighting in each. They knew that rats have a wide repertoire of play-fighting moves, but they found that house mice have only two: attack and run away. One mouse will make a playful attack on the other, who will escape, perhaps returning to counterattack, perhaps not. That's it. If the play of house mice doesn't sound like much fun to you, a rat would probably agree. Still, as sadly impoverished as their play fighting is, it's better than some. Certain hopping mice, poor things, do not play fight at all.

When the Pellises compared play among sixteen types of murids, they found that four (those hopping mice among them) did not play fight. The twelve that did play fight (rats among them) performed at various levels of sophistication. The play fighting of Syrian golden hamsters was among the most sophisticated, with all the protocols and strategies of rats. The play of European voles was among the least sophisticated and, like the play of house mice, limited to attack and run away.

The Pellises mapped the play behavior of each of the twelve playful murid types onto a cladogram and traced them backward stem to branch to larger branch—and finally to an ancestral murid, the great-great-grandfather of them all. The cladogram suggested that he had play fought more than house mice and European voles, but not so much as rats and Syrian golden hamsters. If on the cladogram you traced play behavior from that ancestral murid forward along each branch, you'd see that along some it grew more sophisticated, and along others it disappeared altogether. Natural selection giveth, and natural selection taketh away. The cladogram showed, or at least implied, something else. Since the ancestral murid play fought and some hopping mice don't, somewhere on the branch between them there must have been a rodent for whom the costs of play fighting outweighed its benefits, and for whom play fighting offered no adaptive advantage. This particular animal, the John Calvin of murids, had found that play hath no profit.

Play-fighting murids typically attack and defend the same places that they target and defend in their courtship behavior. Montane voles and prairie voles, like rats, attack and defend the nape. Syrian golden hamsters nibble the cheeks, and northern grasshopper mice lick and nuzzle the shoulder and side of the neck.[2] The undisputed titleholders of murid adorability, Djungarian hamsters, try to lick the mouth of the other, which researchers call "kissing." Most of the play fighting of murids, such as these, is sexual. But some of it is aggressive. Conveniently for researchers, each move in a play-fighting sequence is distinguishable from others in that sequence. An attack on the nape is easily identified as sexual; an attack on a flank, aggressive.[3] The moves of healthy murids are sufficiently discrete that researchers can quantify what proportion of their fighting is sexual, and what proportion is aggressive. Those adorable Djungarian hamsters don't always try to kiss and aren't always adorable. For every ten moves that are sexual, three are aggressive.

Murids, fascinating as they are, comprise only a single suborder of mammal, a division of rodents known as *mouse-like*. The two other divisions of rodents are the *squirrel-like* (among which we count the gray tree squirrel and African ground squirrel) and the *guinea pig–like* (which along with its namesake includes the North American porcupine and the degu). While the play fighting of murids is mostly sexual, the play fighting of the other divisions of rodents is mostly aggressive.

Degus' play fighting, for instance, is much like their real fighting. The targets they attack and defend are the same: the shoulders. The postures are the same, too. They grapple each other with their forepaws, rearing on their hind feet, all the while face-to-face. From this position they can maneuver easily, kicking their opponent with hind feet and knocking it onto its side or back. Recall that the play fights of rats involve postures—the supine defense and pinning with all four limbs— that allow an opponent to gain an advantage. In degu play fights there's no supine defense, no pinning with all four limbs, no self-handicapping at all. When degus play fight, they play fight to win.

What then makes it count as play? Two things. One is the measure of restraint. Suppose a degu knocks its opponent over. In a serious fight, it would almost immediately lunge to bite the opponent's shoulder. But if that same degu is play fighting, either it would stop altogether, or it would stop and stay still, allowing the fallen to regain its footing and launch an attack to keep the play going.[4] The other thing that makes it count as play is that the participants affiliate—or "stay friends"—when the bout is over.

If we zoom out from the cladogram of the evolution of play fighting in murids to take in a larger cladogram that includes the squirrel-like and the guinea pig–like rodents, we'd see something interesting. The playfighting of murids and that of degus have different evolutionary histories and different adaptive advantages. Most of the play-fighting of murids evolved from courtship behavior. One of its adaptive advantages is practice for courtship. Most of the play fighting of degus evolved from real fighting. One of its adaptive advantages is practice for real fighting. Yet in neither case is natural selection quite finished. For both animals, play fighting provides *other* adaptive advantages, too. It helps develop skills in social competence, social negotiation, and social assessment. And it supplies training for the unexpected.

Suppose we wished to construct a larger cladogram. From the cladogram of play in all rodents we'd extend more lines outward. Some would run parallel, others would diverge, some would end abruptly, and still others would branch and branch again, producing the networks that evolutionary biologists call *radiations*. Including more and more paths and ever larger radiations—that of all mammals, then all vertebrates, and finally all animals—we'd eventually derive a map that would yield a greater, more holistic, and far more satisfying understanding of the evolution of animal

play. Such a project would be ambitious, and given the probably many thousands of species of playing animals that evolved
over the last 500 million years, it's unlikely that it could ever be
completed. But it may not be too soon to imagine how to begin.

One might start by comparing the evolutionary history of
rodent play with that of another mammalian order: primates.
The category includes some species we've met already—monkeys
and macaques—and some we'll give more attention to here:
lemurs, gorillas, and ourselves—*Homo sapiens*.

The Complicated Courtship of Pottos

Primates' play fighting is woven into their mating behavior, and
it complicates that behavior in interesting ways. The Pellises
hypothesize that for animals that are by nature independent and
wary of intruders, play fighting is a necessary stage in courtship,
the only viable course to intercourse. This seems to be the case
with pottos, small nocturnal primates native to the tropical forests
of Africa. They are fiercely independent. Male and female pottos forage alone and generally sleep alone, and they may attack
other pottos who invade their territory. But a male potto with
amorous intentions will visit a female whose territories overlap
his own, returning for several nights. If she becomes accustomed
to his presence, she may allow him to groom her, and he may
allow her to groom him as well. Then something happens that
might seem strange. The two pottos, hanging upside down by
their feet near each other, begin play fighting—grabbing and
pulling at each other with their forelimbs. After several nights of
this, they may copulate. Or not. Pottos, it seems, can be moody.

We share an ancestry with pottos, and it would not be surprising to see something of ourselves in them, or them in us. Human mating behavior often involves play. One study noted, "Humor might function as a fitness indicator associated with greater desirability,"[5] thus affirming a truism so obvious that it may not have needed affirming. Human mating behavior also often involves elements of aggression. The on-screen couple who scrap and squabble potto-like through the first eighty-five minutes of a movie and spend the last five in each other's arms is a Hollywood rom-com cliché. And then there's the tango.

Widening the Search

Ethologists have undertaken many hundreds of studies of play in rodents and primates. They've also researched the behavior in wolves, cows, elephants, whales, and dolphins. All are members of an order of mammals called *placental*, that is, mammals having a placenta that develops during gestation. Other placental mammals—such as scaly anteaters, hedgehogs, shrews, and moles—have received less attention, perhaps because they are not as charismatic, perhaps because their play is less than spectacular. To the untrained observer, much of it might look like poking through leaves. Yet such animals have been shown to play to some degree, or at least exhibit something like it— what Burghardt and other ethologists cautiously term "play-like behavior."

All placental mammals are descended from an evolutionary line that is believed to have diverged at least 100 million years ago from another mammalian line, one that produced the order of mammal called marsupials. Unlike placental mammals, marsupials are born not fully developed and are, in many cases, suckled and carried in a pouch on the mother's belly. Among the order's members are kangaroos, wombats, bandicoots, and opossums. Many are known to play, and some play fight in ways that look much like the play fighting of their placental cousins. Play-fighting kangaroos, for instance, adhere to a protocol. One invites another to play fight, the other accepts, and they then begin a series of bouts during which either or both may self-handicap. Their goal is not to harm the other—as would be the case in a real fight—but only to push or wrestle the other off balance.[6]

Refining the Search

Most studies of animal play focus on the play of placental and marsupial mammals. There are roughly sixty-four hundred species of mammals by one recent count, but estimates of the total number of animal species on earth fall within a staggeringly wide range, with some informed guesses putting the number as high as 7.7 million.[7] If among that number we wish to find animals that play, we'd be prudent to find a way to—here borrowing a phrase from another field—limit our search parameters. Burghardt has proposed one. Since play is enabled by certain physiological, behavioral, and developmental conditions, we might look for animals provided with them. He's identified five.

Play calls upon a surfeit of energy above that which an animal requires to merely stay alive. An animal that has a surfeit of energy is likely to have a high metabolic rate. Most animals with high metabolic rates can generate heat from within their own bodies and are colloquially termed *warm-blooded* or, more formally, *endothermic*. An animal that plays, says Burghardt, is likely to be endothermic.

To sustain said metabolic rate, an animal needs sufficient calories. An animal that plays is likely to be one that has a nutritious and plentiful diet.

Animals that are hungry or feel threatened play less than they do when relaxed, or not at all. But an animal will not play if it is *too* relaxed. Thus an animal that plays is likely to be relaxed, yet seeking stimulation.

If primary process play—that fidgeting, squirming, and wriggling—is to develop into more complex, sophisticated play, the animal needs time to develop it. Thus an animal that plays likely has a long juvenile period.

An animal that invents play moves does not invent them from scratch, but borrows them from other behaviors and adapts them for play. Such an animal has behavioral flexibility, and the more behaviors it possesses, the more it has available to adapt. Thus an animal that plays likely possesses both behavioral flexibility and a wide behavioral repertoire.

Looking Where the Light Is—and Isn't

Burghardt's list of conditions helps us make guesses as to which animals play, which don't, and why. For instance, all other things being equal, an animal with a higher metabolic rate is more likely to play than an animal with a lower one, and an animal with a nutritious and plentiful diet is more likely to play than an animal without one. Since both mammals and birds have an aerobic metabolism that enables vigorous activity, and both practice and benefit from parental care, it's not surprising that members of both classes play, as different as they are. The list also enables us to make finer discriminations, to predict not only the presence of play, but also its sophistication. Knowing, for instance, that an animal that plays is likely to have had a long juvenile period, we might have predicted that rats, whose juvenile period is twice as long as that of mice, would have far more sophisticated play. And they do.

That said, all things are seldom equal, and Burghardt does not intend his list as definitive. To use it as our *only* guide, to look for play only in animals afforded these conditions, we'd be acting a bit like the man looking for car keys under a streetlight because, as the joke goes, "This is where the light is." If, for instance, we limited our search to endothermic animals with high metabolisms, we'd neglect turtles.

Reptile Recreation

Walking along a riverbank, I've sometimes noticed what I thought was a knot of driftwood, then realized it was a box turtle sitting perfectly still. Waiting for it to move, anyone would suspect that the species, and perhaps reptiles generally, are simply too solitary, too incurious, and too sluggish to play.

Through much of the twentieth century, many ethologists held similar views, and thought that play was found exclusively in mammals.[8] They dismissed what looked like play from nonmammals either as misfiring instincts or as fragments of developmental processes. Such skepticism is understandable. After all, reptiles are ectotherms, without the high metabolic rate that Burghardt finds necessary for play. But reptiles are descended from the same small, bipedal dinosaurs that gave rise to birds. This shared ancestry has prompted some evolutionary biologists to argue that the traditional phylogenetic categories should be reshuffled and renamed, that birds are properly categorized as reptiles, and that box turtles—along with crocodiles, lizards, and their relatives—should be termed non-avian reptiles.

Some reptiles enjoy some of the conditions that Burghardt contends enable play in birds: plentiful and nutritious diets, extended parental care, and a diverse behavioral repertoire. Although most reptiles are endothermic, some may find ways to produce relatively high metabolic rates at least for brief periods, by, for instance, warming themselves in the sun. Given all this, and realizing that there are thousands of reptile species—crocodiles, alligators and gavials, worm lizards, turtles and tortoises, snakes and lizards—we might conjecture that at least some of them play. Many ethologists have harbored such suspicions, but few have acted on them.

One reason is that identifying play in reptiles, like looking for keys in the dark, is difficult. Solitary play is easy enough to see in animals like that frolicking pony from several chapters back, but it's more difficult to recognize in birds, and far more difficult to recognize in reptiles, animals whose movements are slow and deliberative, and whose exuberance, if they have such, is well contained. Thus the challenge. Ethologists who look for play in reptiles must be open-minded enough to recognize it when it's there, yet not so open-minded that they imagine it's there when it isn't. They need an unambiguous and straightforward definition, and many have made use of Burghardt's, introduced in chapter 1 and in chapter 7 applied to dreaming, and worth quickly restating. Per Burghardt, play is behavior that is nonfunctional, voluntary, characterized by repeated but varied movements, and performed by an animal that is well-fed, safe, and healthy.[9]

Some anecdotal accounts of reptile behavior meet the definition, with examples from each of the traditional play categories. A captive wood turtle sliding down a board into water, slowly climbing up the board, and sliding down again. A juvenile loggerhead turtle performing a marine version of herring gull drop-catching—swimming beneath the surface, releasing a semibuoyant plastic ring into a rising column of water, catching it, and releasing it again. A particularly remarkable account of play is of a turtle that shared a tank with a nurse shark. On occasion, the turtle would gently bite the shark's tail. When the shark tried to swim away, the turtle would hold on for a few moments, letting itself be pulled along by the shark, and perhaps taking pleasure in a free ride.[10]

Several rigorous studies of reptile recreation have focused on turtles. One such study, undertaken by Burghardt and two colleagues, involved a rather sizable resident of the Washington National Zoo, a Nile softshell turtle weighing nearly sixty pounds.[11] The researchers could not be sure that the turtle practiced solitary play, but they noted that it often swam not to get anywhere but, it seemed, just to swim. Since turtles are foragers who explore and investigate, the team expected that their subject might readily engage in object play. To prompt it, they introduced several articles they thought the turtle might regard as toys—a basketball, some large sticks, and a hoop of rubber hose. The turtle nudged, bit, chewed, and otherwise manipulated the basketball and sticks and sometimes swam through the hoop. Like Octopuses 7 and 8, those putatively playful cephalopod test subjects of Mather and Anderson, the turtle seemed not merely to be investigating the objects, but trying to find out what it could do with them. Burghardt and colleagues concluded that the behavior could be counted as object play. The turtle also engaged in social play, not with members of its own species (perhaps because none were available), but with its human keeper. The turtle would nose, claw, and bite the hose the keeper used to replenish water in its tank. When the keeper pulled the hose while the turtle had one end in its mouth, the turtle, as though playing tug-of-war, would bite the hose harder and swim backward.

A Playful Komodo Monitor Lizard

Burghardt investigated play in another solitary resident of the zoo, a female Komodo monitor lizard that its caretakers had named Kracken,[12] a fitting honorific for this member of a species weighing as much as three hundred pounds and reaching lengths of ten feet, and whose preferred meals are goats, wild boars, and deer.[13] For all their ferocity, monitor lizards are known to play. And like the turtle, Kracken nudged and bit various objects, among them a Frisbee and a tennis shoe, in a manner that seemed to go beyond investigation into play.

Monitor lizards have the highest standard metabolic rate of all reptiles, yet like turtles they are ectotherms. To a human (that is to say, an endothermic) observer, their play may seem a bit sedate. But Burghardt found that what counts as sedate for an endotherm may not be sedate for an ectotherm. Over two years, he and his colleagues made several hours of video recordings of both the turtle and the Komodo lizard. When they watched the recordings played at higher speeds, they found the animals' movements to be surprisingly familiar. When Kracken bit and shook the tennis shoe, Burghardt thought, she looked a lot like a dog.

Insofar as Burghardt's list of conditions that enable play is a rule, then turtle play and Komodo lizard play are one of natural selection's work-arounds—and further cause to admire and perhaps be astonished by the ingenuity of which it is capable. They are also further evidence that the adaptive advantages of play, whatever they are, are significant.

Examples of reptilian recreation invite us to wonder exactly how unlike a mammal, how far from the branch of the phylogenic tree bearing mammals an animal can be, and still play. Until recently, most ichthyologists considered fish play unlikely, if they considered it at all. But evidence is accumulating that fish cooperate, imitate, and even deceive. So we might reasonably ask, Is there fish fun? Herring high jinks? It would be imprudent to generalize. There are some thirty thousand species of fish, with an impressively wide range of metabolisms, degree of parental care, and behavior—a range as great as that of mammals. And many fish species are afforded at least some of the conditions that enable play in mammals.

Piscine Play

One such condition is a high metabolic rate. Since most fish are ectothermic, we might suppose they lack the surplus energy necessary to play. However, the energy an animal requires to move varies greatly depending upon the medium on, over, or through which it moves. A bird flying a given distance through air expends less energy than does a mouse of the same weight scurrying that distance on the ground. A fish swimming that distance expends less energy than either. If the fish isn't particular about where it moves, it doesn't need to expend much energy at all. For most purposes it's weightless, and with a slight stroke of its fins or tail it can enter a water column and be carried a great distance upward or enter a current and be carried a great distance horizontally. Arriving at a place it doesn't like, it can, with as little effort, set off again. So if a fish's metabolism is not high enough to play on land, it may be more than high enough to play in water.[14]

Fish may engage in a relaxed sort of solitary play, like the turtle in the National Zoo, swimming just to swim. Fish of many species have been seen to dart about alone, undertaking rapid dives as though chasing or being chased. A somewhat dated study speculated that they are reacting to imaginary playmates.[15] Individuals in many species behave in a way that looks like teasing. One fish incites another, then darts away, and for a few seconds the other chases it. Groups of two or three young fish have been seen to suddenly leave the larger school in which they have been swimming, chase each other for a few moments, then return. In what looks like "playing catch," two gourami were reportedly seen to spit recently spawned eggs from one to the other and back again.[16] Captive sturgeon in the Aquarium of the Pacific in Long Beach, California, were reported to swim to the surface, gulp air, descend to the tank's bottom, release the air as bubbles, and then chase them upward.[17] None of these behaviors have been studied at length, and each may well be something other than play.

Some fish, though, do something quite likely to be recreational: leaping. That is, jumping out of the water and plunging back in. Not all leaping is play. When flying fish leap and use their winglike pectoral fins to glide long distances above the water's surface, they are escaping predators. Some fish leap so that when they fall back, the impact of their skin against the water's surface dislodges parasites. Fish living in water with particularly low levels of oxygen may leap to momentarily expose gill filaments to air, leaping, as it were, to breathe. Yet needlefish, halfbeaks, herring, and silverside minnows have reliably been reported to leap for none of these reasons. All have been observed leaping and—this is important to note—not just once and not just anywhere, but repeatedly and over something floating on the water's surface— sticks, pieces of straw, and in one case a sleeping hawksbill turtle. More impressive still, several reports say a species of halfbeaks undertakes bodily pole-vaulting. The fish approach a floating object, push their jaw against it, and swing their tail upward and forward, pivoting on the jaw and flying out of the water tail first, over the object, striking the water beyond.[18]

In one remarkable account, fish engaged in social play with a member of our species. A worker at the Senckenberg Research Institute and Natural History Museum who fed redeye and rudd coaxed them to rest in his hand as he held it near the water's surface. The worker then began to lift the fish from the water and toss them back in. For the fish, the experience was evidently pleasurable, since most swam back to the worker's hand and allowed themselves to be tossed again.[19]

The late behavioral ecologist Erich Ritter experimented with a chunk of frozen fish and a large floating ball hung over the side of a boat. Sharks fed on the fish, and some pushed the ball around. When the fish was pulled back aboard the boat, the sharks continued to push the ball. If they came for the food, they stayed for the play. And they seemed to know it carried risks. Sharks attack one another by biting the other's gills; they protect themselves against such attacks by squeezing their gill slits together. The sharks Ritter observed pushing the ball had closed their gill slits. This meant, he thought, that they knew they were risking injury to play, and they were playing nonetheless.[20]

Mammals, birds, reptiles, and fish are vertebrates, and their brains, as different as they are, share a lineage. All vertebrate brains are variations on the same model. Comparative neuroscientists can identify parts of a human brain that correspond to a part in a rat's brain, a bird's brain, a reptile's brain, and even a fish's brain. The brains of octopuses, as with so much else about them, are different. Yet, as we've seen, octopuses play, although perhaps not our games, and not by our rules. Might play also be found in another class of invertebrate?

Bees Do It

Insects comprise a class that we might suppose unlikely candidates for play—their brains too small, their nervous systems too simple, their behavior too rigid. Such presuppositions were given weight by preeminent entomologists Bert Hölldobler and Edward O. Wilson. In their 1990 landmark study, *The Ants*, they wrote, "We know of no behavior in ants or in any other social insect that can be construed as play or any other social practice behavior approaching the mammalian type."[21]

But then there are bees. We have known for decades that bees work collectively, and that they use symbolic language in the "waggle" dance with which they communicate the location of food. More recently we've learned that bees can follow intricate rules, distinguish patterns, and differentiate shapes and colors. We've also learned that they are endowed with rudimentary mathematical ability: they can count.

In 2022 a team of scientists at Queen Mary University of London performed a series of experiments to determine whether bumblebees played.[22] They provided forty-five bees a choice: an area through which the bees might walk to receive food, and another strewn with bee-size wooden balls. Over eighteen days, they recorded 910 instances during which bees rolled a ball. Some rolled balls only once; others did it more than forty times in a single day. After considering possibilities that the behavior might be exploration for food, clearing clutter, or mating—the team dismissed all as implausible and concluded that it could only be play. As they noted, it fulfilled Burghardt's criteria: it was nonfunctional, voluntary, characterized by repeated but varied movements, and undertaken when the bees were well-fed, safe, and healthy. Moreover, they found additional evidence for their conclusion in two features of the bees' behavior that resembled the play of mammals: males played more than females, and younger bees played more than older ones.

More than a million species of insects have been identified and documented, and all evidence suggests the total number of insect species on earth is far greater. Yet the work of the researchers at Queen Mary University of London is one of a very few serious attempts to identify play in insects.[23] Given how little we know of insects generally and insect behavior specifically, it hardly follows there is none.

Ideas of Play's Beginnings

Play may have begun independently in a great many evolutionary lines, arising in the long history of life on earth at different moments in the evolution of many animal species. Alternately, it may have arisen once and branched and branched again across phyla and classes.

During the Ediacaran period, 543 million years ago, earth's landmasses were barren. What life there was lived in the oceans. A film of microbes covered the seafloor, and most animals were sponges or slow-moving worms. Existence was peaceful and unhurried. Nothing moved quickly, and most living things didn't move at all. Then, some 541 million years ago, that long quiescence ended with a diversification of life forms so dramatic and wide reaching that it's come to be known as the Cambrian explosion. The oceans teemed with life. There were trilobites, thousands of species. Brachiopods with clam-like shells. Worms with feathery gills. Centipede-like creatures with compound eyes. Streamlined, fast-swimming predators—the forbears of fish and eels—that could seize prey in tooth-rimmed jaws. And cephalopods, one with an eighteen-foot-long cone-like shell and a head of muscular tissue from which it extended tentacles.

Many of these animals had brains and nervous systems, and some may have reached the level of sophistication necessary for play. We don't know and most probably never will know which animal was the first to play. Nonetheless, it's tempting, and perhaps instructive, to imagine one. It may have been the *Myllokunmingia*, a member of the phylum Chordata that lived in the lower Cambrian, some 518 million years ago. Paleontologists have found only a single specimen, a fossil buried in sediment in Yunnan Province, China. It's about the size of a guppy, with skeletal structures of cartilage, a distinct head, and a trunk with a forward sail-like dorsal fin—all features that suggests it was free swimming. If *Myllokun-mingia* played, we can be reasonably sure that its play would have been of the most rudimentary, primary process sort, a by-product of "excess metabolic energy," and little more than a twitch or wriggle.

Or, the first animal to play may have been a member of the phylum Mollusca. Fossils of one were discovered on the Avalon Peninsula in Newfoundland. A probable ancestor of the present-day nautilus, it lived 522 million years ago. Like the nautilus, it did not swim, but adjusted its overall buoyancy by moving fluid in and out of a long cone divided into chambers.[24] Its play, like that of *Myllokunmingia*, would have been simple—perhaps just letting itself be carried along by a gentle current or lifted in a rising water column.

Just as it's possible to imagine many animals that were first to play, it's possible to imagine many routes by which play evolved and diversified. It may have appeared once during the Cambrian period when conditions were right, branched and branched again, disappearing in some lines and reappearing in others. Or it may have appeared many times along many evolutionary lines, in some developing only as far as primary process play, in some to secondary process play, and in some still further to tertiary process play and its countless manifestations: the ball-bouncing of octopuses, the play fights of kangaroos, the drop catches of Montagu's harriers, and the dulcet trill of a wood thrush.

A Cladogram of All Play

Recall the conditions that Burghardt suggests enable play: sufficient metabolic energy, a comfortable and stress-free existence, a need for stimulation, a long juvenile period, behavioral flexibility, and a varied behavioral repertoire. Burghardt posits that they may provide the basis for predicting the presence of play in a given animal. If all are present, that animal will play; if none or only some are present, that animal won't play. Since these conditions are present in degrees, with more pronounced conditions compensating for some that are less pronounced, we might think of them as a set of slider controls with adjustable settings. So long as the average of all the settings falls within a given range, the conditions enabling play are met.

One might test the hypothesis across a radiation. In some species we'd find that play appeared as predicted, confirming our hypothesis. In other species it might not appear as predicted, or perhaps, as in the case of octopuses, it did appear when it *wasn't* predicted.

In fact, were we to use Burghardt's conditions as our only guide to animal play, we might take a pass on octopuses. Most species live for only three or four years, and none have a long juvenile period. How, we might wonder, do they have time to learn to play? Burghardt's answer is simply that longevity is relative. He notes, "A short-lived species that breeds three times in one year must have its developmental period calibrated against a species that may only reproduce once every two years."[25]

Conditions enabling play that octopuses lack may be compensated for by others that are abundantly present. Octopuses are well endowed with behavioral flexibility and a wide behavioral repertoire, both to be expected in animals that seek varied food sources in changeable, perilous habitats. Jennifer Mather notes that any creature "foraging in a complex environment, where you have to deal with many kinds of prey and predators," must develop a wide range of hunting and defensive strategies.[26] It should not be surprising that octopuses are especially adept at object play.

Upon identifying play where it wasn't predicted, we'd need to reevaluate, to tease the conditions apart, weigh them against one another, readjust the slider controls as necessary, and test again.

How to predict the path by which the animal's play evolved *after* that appearance? A more challenging prospect to be sure, but the inflection points of Burghardt's surplus resource theory might offer a way to begin. The moment those points appeared would depend upon certain features. The secondary process play of the ancestor of capuchin monkeys, for instance, might have been made possible by the development of a sufficiently sophisticated neural circuitry. Identify the moment when the capuchin ancestor developed sophisticated neural circuitry, and you might have identified an inflection point in the evolutionary history of its play.

Genes and Master Genes

In *On the Origin of Species*, Darwin acknowledged that several parts of his theory presented "difficulties," one of which was that it did not identify a mechanism by which natural selection operated. He would never learn that in 1865, six years after the publication of *Origin*, Moravian monk Gregor Mendel posited such a mechanism, demonstrating that characteristics are transmitted from parents to offspring by the units of heredity that would later become known as genes. In the twentieth century, evolutionary biologists came to understand genes as a distinct sequence of nucleotides forming part of a chromosome. The scientists also confirmed Mendel's finding that genes are specialists. In fashioning the eye, for instance, one gene provides information to construct a light-sensitive pigment, another information to make a lens, and so on. Evolutionary biologists have long believed that eyes, as products of natural selection, owe much to happenstance, since the assembling of the genes necessary to create eyes is a mostly random trial-and-error process taking place over extended time. In the "credit where credit is due" category, it turns out that those genes may have had some help.

In the 1970s scientists brought molecular genetics—that is, the study of heredity and variation at the molecular level—to bear on embryology, the study of embryos and their development. The result was a scientific field termed *evolutionary developmental biology*, or more informally *evo-devo*. Its practitioners are learning how embryonic development is controlled at the molecular level, how the development of an individual relates to its evolutionary history, and how developmental processes themselves evolve.

Already they've met with two surprises. The first is that not all genes are specialists. Some genes—*master genes*—are generalists, higher-level administrators that trigger the activation of ordinary genes and coordinate their functions. The second is that master genes are "highly conserved" over time, meaning that while a given feature has evolved over hundreds of millions of years, the master genes that trigger the ordinary genes shaping that feature have not.

Master genes explain some cases of what's called *convergent evolution*. The common ancestor of octopuses and humans either had small patches of skin that were sensitive to light or was entirely blind. Darwinian evolution, confronted with the same challenge—how to detect prey and predators at a distance— arrived more than once at the same answer. Octopus eyes and our own are both equipped with an adjustable lens that focuses an image on a retina. Risking cuteness in the service of a rhymed couplet, we might call it an adaptive advantage so useful and nice that natural selection selected it twice.

In 2014 scientists at the Nagahama Institute of Bio-Science and Technology in Japan found that the similarity of the eyes across animal species is owed to a shared master gene. More than 500 million years ago, a master gene termed *PAX6* orchestrated the formation of a patch of light-sensing cells belonging to an animal living in the Cambrian seas.[27] In the time since, PAX6 has not changed, but its responsibilities have become far greater and more diverse. It now controls the development of all manner of eyes—the eyes of humans and octopuses, as well as the compound eyes of leaf-cutter ants, and the telescopic eyes of red-tailed hawks.

The Role of Genes in the Evolution of Play

Genes determine physical features—such as eyes. Play, though, is not a physical feature. It's a behavior. Although genes can't determine a behavior directly, they can determine the trait or set of traits that make a behavior possible—in the instance of play, perhaps a certain neural circuitry. If in the evolution of animal play there were many first appearances, each may be owed to new mutations or new ordinary genes. But it's also possible that each is owed to the same set of master genes lying in wait along several evolutionary paths, and when an animal is provided the conditions that make play possible, those master genes are stirred. They activate ordinary genes; those activate still other ordinary genes, and some produce the physical features in an animal—as for instance that certain neural circuitry—that enable play.

At present, evolutionary biologists do not know that a master gene enabled and orchestrated play, much less *which* master gene. Neither do they know where or when play began. They have no map, no cladogram, depicting the evolution of all animal play. But they know that play has a history stretching back hundreds of millions of years, and that its roots, that hypothetical suite of master genes, may be older still. Play has endured the formation and reformation of continents, three ice ages, and two mass extinctions. So they—and we—can be certain of one aspect of play. Whatever its adaptive advantages, they are worth the trouble. Nature takes play seriously.

In this chapter we've shown how natural selection influences play. In the next we'll consider how play influences natural selection.

CHAPTER 9

Innovative Gorillas:
The Surprising Role of Play
in Natural Selection

While many play-fighting animals—rats and degus for instance—always attack and defend the same body parts, gorillas have been known to seek alternatives. Their usual target is the shoulder, but on occasion they may prefer to mouth or gently bite another's foot. The Pellises found that some may venture still further afield, noting, "After repeated attacks on the shoulder, a juvenile grabbed the crotch of another juvenile with his hands. The recipient's startled response and subsequent leap into the air suggests that it was as surprised as we were!"[1]

Other primates also innovate their play, and their originality can catch researchers off guard. When Milada Petrů's team studied those five species of monkey, they saw individual animals routinely engage in the same sort of play, so routinely that they weren't particularly surprised to see another vervet monkey performing another somersault, or to see another patas monkey practicing parkour. They were surprised, though, when a de Brazza's monkey turned around on all four limbs to complete a 360-degree rotation, and when a Diana monkey lifted an object above its head and hopped in place. Both monkeys were playing in a way unlike any other in their group, a way that was singular, idiosyncratic, and innovative. If that crotch-grabbing juvenile gorilla was changing the rules of old and well-established games—acting, in the Pellises' phrase, as a "theme-breaker"—the monkeys were inventing new ones from scratch. They were—here again a Pellis coinage—"playing with play."

What enables primates to innovate? One might suspect it's their brains. Primate brains are considerably larger than those of other mammals both in absolute terms and relative to body weight. Generally, mammalian orders with larger-brained members tend to play more, with humans being the most playful. But in smaller taxonomic units, such as families within an order, the rule does not hold. The Pellises suspect that primates' innovation is likely owed not to their brains, but to their child-rearing practices. Those practices may be best illustrated in the behavior of a small arboreal and nocturnal primate that inhabits the forests of Sri Lanka, India, and Southeast Asia—the loris.

Loris Play

A mother loris usually gives birth to one offspring at a time, rarely two. Most evenings she covers her infant in allergenic saliva, a toxin that discourages predators. She then takes care to hide it in a locale she believes to be safe and leaves to spend the night foraging. Thus the infant is alone for long periods and has only one occasional playmate—its mother. The loris of our own time is descended from a distant ancestor whose lineage would branch again and again, producing lines that would eventually lead to all extant primates: the great apes, monkeys, macaques, and us. Primatologists suspect that the mother-infant play of that ancestor has endured, little changed, in the loris of our day. They suspect that it's also persisted in other primate lineages. It's quite evident, for instance, in gorillas. Gorilla mothers don't just nurse, protect, cuddle, and groom their infants. They also play with them. They swing them, engage them in peekaboo, and, as many a delighted human observer can attest, treat them to the kinetic pleasures of "airplane." The mother lies on her back, supporting and balancing an infant with an upstretched leg, her foot pressed against the infant's chest to provide a measure of stability, the other foot holding one of the infant's feet and gently steering him.

Most mammals first experience extended periods of play as juveniles and with other juveniles. But the gorilla, like many primates, gets a kind of play preschool in infancy, learning from its mother what it feels like to be jostled, poked, and thrown off-balance. When it becomes a juvenile and begins to play with peers, it is prepared to appreciate nuances and, perhaps most significantly, to experiment and innovate.

The innovations of primate play, the "playing with play," is a thread in the larger narrative of animal play. It might also be like a thread on the inside of a glove. You pull it only to find that it's not loose and won't break. You keep pulling until you find that, without meaning to, you've turned the glove inside out. So it is, or may be, with innovation in play—the thread being innovation, and the glove nothing less than the theory that underpins our understanding of all life on earth: natural selection. Several ethologists, some of whom we've already discussed, believe that innovative play might be a means by which an animal gains a measure of control over its own evolution.

The Hypotheses (and Controversies) of Evolution

By the second half of the nineteenth century, many scientists, perhaps most, had accepted that organisms had developed and diversified over the history of the earth. But precisely how they had done so was a matter of no small controversy—or rather controversies, since four or five major schools each championed a different hypothesis. The orthogeneticists held that laws of development and organisms' internal forces drove evolution in particular directions. Mutationists argued that evolution was by and large the product of mutations creating new forms or . species in a single step. Of the two groups best known today, the Lamarckians proposed that acquired characteristics could be inherited, and the Darwinists posited that species evolved because organisms better adapted to their environment tended to survive and produce more offspring.

The debate was never long at rest. As new evidence emerged and older evidence was discredited, some shifted their positions. Harvard botanist Asa Gray (who worked to reconcile natural selection with Christianity, reckoning God as the source of all. evolutionary change) likened the controversies themselves to natural selection. They were part of a process, he said, "through which the views most favored by facts will be developed and tested . . . the weaker ones destroyed . . . and the strongest in the long run alone survive."[2]

As if this weren't complicated enough, the names for these schools changed over time and differed from place to place. Since Darwin himself allowed for Lamarckian evolution, he was not, strictly speaking, a Darwinist. Among those who *did* count themselves Darwinists were Alfred Russel Wallace, whose independent discovery of the principle of natural selection prompted Darwin to publish *Origin* earlier than he had planned; the English biologist Thomas Henry Huxley, Darwin's staunchest and most combative advocate; and Charles Lyell, from whose 1830 *Principles of Geology* Darwin had drawn his appreciation for the vast geological timescales his theory required. Each differed regarding certain aspects of natural selection as it had been presented in *Origin* and its sequel, the 1871 *The Descent of Man, and Selection in Relation to Sex*. Lyell, for instance, held that higher faculties in humans seemed a product of divine intervention; he could not, in his own memorable phrase, "go the whole orang."[3] But all held natural selection to be the main driver of evolution.

Nowadays the theory of natural selection is much alluded to, but so often misunderstood, that it's worth a brief summary here.

Natural Selection

The theory follows from two observations, both widely accepted as self-evident. The first is that since more individuals of a given species are born than resources can possibly support, they must compete for those resources. The second is that characteristics of an individual are passed from parents to offspring, and some of those characteristics vary from those of its parents. While most variations are of little consequence and others are detrimental, some offer an advantage. An individual with an advantageous variation is "more fit," that is, better adapted to its environment, and so more likely than those lacking such a variation to survive long enough to reproduce. If that individual does reproduce, it may pass on its advantageous variation to its offspring. Some of those offspring may exhibit still other advantageous variations, add these to those that they inherited, and pass both sets on to their offspring. And so on to the next generation and the next. In Darwin's words, "Any variation, however slight and from whatever cause proceeding, if it be in any degree profitable to an individual of any species . . . will tend to the preservation of that individual, and will generally be inherited by its offspring."[4]

The process described by Darwin is today recognized as, if not the only driver of evolution, certainly the principal one.[5] Yet in 1896, the year Groos published *The Play of Animals*, nearly forty years after the publication of *On the Origin of Species*, the theory's ultimate triumph was hardly foregone. As Groos observed, "The conception of evolution itself is gaining strength and assurance with the progress of time, but with respect to specific Darwinism a note of fin-de-siècle lightness is audible to the attentive ear."[6]

The Neo-Lamarckians

For all the defenses of natural selection that the Darwinists (and Darwin himself) could mount, they were well matched by several competing theories of evolution and associated schools. Perhaps the most formidable and outspoken of the Darwinists' competitors was a loosely aligned group who called themselves neo-Lamarckians, after the French naturalist Jean-Baptiste Lamarck. Lamarck was a deist and held that God had established a divine plan at the creation and did not intervene thereafter. In his 1809 work *Philosophie Zoologique*, Lamarck proposed that this plan was put into motion by evolution, a process that he believed to be decidedly progressive, with organisms over time advancing upward on a metaphorical ladder of phyla through ever-more complex forms, from worms to mollusks to vertebrates. His theory also asserted that organisms adjusted to various environments and grew more diverse by developing adaptations in their lifetime and passing those adaptations on to their offspring. In Lamarck's well-known example, giraffes, reaching for leaves in trees' upper branches, stretch their necks and forelegs, thus acquiring characteristics—longer necks and forelegs—that their offspring inherited. Darwinists would say that a population of giraffes over many generations would exhibit many variations, only some of which would be advantageous. Among the advantageous variations would be longer necks and forelegs, and subsequent generations would show more individuals with them.

Many current textbooks suggest that little or no evidence supports the claims of Lamarck and neo-Lamarckians. This rendition of history appeals to our desire for simplicity, but it is, alas, false. Gaps in the fossil record suggest rapid changes consistent with Lamarckian evolution, much at odds with the gradual and incremental changes posited by the Darwinists.[7] Textbooks today also err in suggesting that neo-Lamarckians were naïve. Little could be further from the truth. They were well regarded as scientists, counting among their number American paleontologist Edward Drinker Cope, British botanist George Henslow, and German zoologist Theodor Eimer. Their position gained a modicum of support from Darwin himself. In *Origin* he noted, "I am convinced that Natural Selection has been the main but not exclusive means of modification,"[8] and although he made extensive revisions to each of the five editions that followed the first, he invariably included a passage that allowed for cases in which acquired characteristics or their absence (e.g., "the wingless condition of so many Madeira beetles")[9] might be inherited.

In the final decades of the nineteenth century, the neo-Lamarckians could answer the Darwinists case by case, example by counterexample. Yet their theory, as they well knew, was missing an important piece. It did not identify the precise means by which acquired characteristics might be inherited. Then, in 1893, German biologist August Weismann claimed that there was none, and none was possible.[10] Complex, multicellular organisms, said Weismann, have two types of cells: reproductive cells and somatic cells. Reproductive cells can pass hereditary information to offspring, but they cannot acquire characteristics. Somatic cells can acquire characteristics, but they cannot pass hereditary information to offspring. Moreover, what would come to be called a Weismann barrier—rather more like a gulf than a wall—prevented somatic cells from transferring characteristics they acquire (or information about those characteristics) to reproductive cells. The Weismann barrier meant that a complex, multicellular organism could not pass characteristics it acquired in its lifetime on to its offspring nor could such an organism inherit characteristics from its parents that they acquired in *their* lifetimes.

The Nature of Nature

While Weismann's work was generally well regarded, the neo-Lamarckians resisted its conclusion not only because it rejected their contention that acquired characteristics could be inherited, but also because it threatened a conception of nature on which their theory depended. In the mid-nineteenth century several scientists hypothesizing about evolution had held that God had a direct hand in it. By the 1890s, though, most agreed that scientific inquiry, even summoning all its resources of hypotheses, experiments, and analysis, could not establish the existence, much less the character, of God. Still, if science could not reveal a Designer, many expected that it might yet reveal a Design. The neo-Lamarckian theory of evolution and the theories of the Darwinists, mutationists, and orthogeneticists each came with a kind of metaphysical underpinning, a set of beliefs and presuppositions about that Design—what we might call the nature of Nature.

Darwin's idea of the nature of Nature took shape gradually, as he assembled the evidence and organized the thinking that would eventually produce *Origin*. Darwin called the book a "long argument."[11] As such, it's a tour de force. It's also a tour of much else, including the cell-making instinct of bees, the flora of New Zealand, seeds that drifted to coral islands, and the sudden extinction of trilobites. In the final pages, as though arriving at a mountain's summit, Darwin seems to pause to catch his breath and take in the prospect surrounding him. As if realizing it for the first time, he observes, "There is grandeur in this view of life."[12] Most would agree. But it was not the sort of grandeur the neo-Lamarckians would have preferred. They found Darwin's theory unsettling for two reasons. One was that the evolution it described included (in Darwin's own words) "no necessary and universal law of advancement or development,"[13] but instead was utterly provisional, responding to conditions of a given place and time. The neo-Lamarckians were equally troubled by its implication that nature was deterministic. An individual animal, let's say a member of our own species, may believe that she has free will, that her thoughts and actions are, or at least can be, independent of forces outside her. Natural selection implied otherwise—that all her thoughts, profound and trivial, and all her actions, effectual and ineffectual, are ultimately driven and shaped by, and only by, the needs to survive and reproduce.

In the last decade of the nineteenth century, one naturalist proposed a modest proviso and annotation to Darwin's theory, suggesting that natural selection and free will were not at odds—at least not always.

Baldwin's Organic Selection

The English-language edition of Groos's *Play of Animals*, edited and prefaced by American scholar James Mark Baldwin, appeared in 1898, two years after the original German. It was Baldwin's preface that first attached the word *practice* to Groos's hypothesis.[14] Owing to Baldwin's influence and reputation, the usage quickly became widely accepted and preferred. Groos could hardly have asked for a better imprimatur. In time Baldwin's "step-wise" theory of cognitive development, which posited that a child learns behaviors in discrete and identifiable steps, would be adapted by Swiss psychologist Jean Piaget and developed into what nowadays is better known as the stage theory. But even in 1898, when Baldwin was only thirty-seven, an age when many in academe are still finding their way, he was already regarded as an authority in the emerging fields of mental development in children, developmental psychology, and evolutionary psychology. He had served as president of the American Psychological Association, authored books on a range of subjects, cofounded the *Psychological Review*, and established two of the first laboratories in experimental psychology—one at the University of Toronto and the other at Princeton, where he held a named professorship.

Baldwin was, in the words of one biographer, "a rather prickly character . . . arrogant [and] much given to controversy."[15] Yet none of this is evident in Baldwin's preface to Groos's work. It did what most prefaces do, supplying a brief chapter-by-chapter summary and capping it off with praise. Groos returned the favors. His book included an entire appendix by Baldwin and endorsed his older work in footnotes. Clearly, the two men were deeply engaged with each other's interests, helping the other to understand aspects of human and animal behavior and their role in instinct, imitation, and, most especially, in evolution.

Baldwin thought little of neo-Lamarckism. He called the theory speculative and noted there was no evidence that characteristics an animal acquired in its lifetime could be passed on or inherited, and no conceivable mechanism by which neo-Lamarckian evolution could operate.[16] Although Baldwin was persuaded by Darwin's theory—indeed he was an advocate—he did not accept the proposition that it precluded progress and free will in every case. In 1896 he argued that there were—must be—instances during which an animal could liberate itself from the forces of natural selection and, perhaps more surprisingly, in some measure actually direct them.

He expanded on this hypothesis in the preface to *The Play of Animals*. (In a sort of convergent evolution of thought, two others, American paleontologist Henry Fairfield Osborn and British ethologist C. Lloyd Morgan, proposed much the same idea, independently of each other and of Baldwin, in the same year.) Baldwin called the process *organic selection*.

Suppose we were present to witness the beginnings of a particular behavior now common in two animals. In a colony of capuchin monkeys, we see one monkey cracking open nuts using two stones, one as a hammer, the other as an anvil. Suppose also that in a population of eastern cottontail rabbits, we see one rabbit escaping predators by running in a zigzag that those predators, unable to change directions as quickly, find difficult to follow. Both behaviors are what the neo-Lamarckians called acquired characteristics (in this case behavioral, not physical), what Baldwin called adaptive modifications, and what geneticists today call *phenotypes*. We leave the capuchins and rabbits unobserved for several generations, and upon returning we find all the capuchins in the colony opening nuts with stone hammers and stone anvils, and all the rabbits in the population escaping predators by running in zigzags. Moreover, we find that both the capuchins and rabbits have increased their numbers.

What happened?

Both behaviors clearly offered an adaptive advantage, one giving the capuchins access to a food source and the other preventing the rabbits from becoming one. We don't know how the first capuchin learned to open nuts with stones or how the first rabbit learned to run in zigzags, and for our purposes it doesn't matter. What matters here is that both are complex behaviors unlikely to be inheritable. Let's assume they aren't. If the increase in population of both animals was attributable only to natural selection, then individual animals would acquire the behaviors by trial and error, and whatever physical or behavioral features enabled them to learn the behaviors would have been selected for. If the behaviors were also learned and imitated by others, then those behaviors would become more common more quickly. In either scenario (or a combination), the increase in numbers would have required a great many generations and would have come at considerable cost, since in each generation many or most individuals that were unable to learn the behaviors would be culled.

Let's hit replay and, this time, imagine the same evolutionary development had it involved Baldwin's *organic* selection. First, consider that all capuchins and rabbits are not created equal. Some are endowed with inherited features that enable them to learn and imitate more easily than others. These features might be physical, such as a specialized neural circuitry. Or they might be behavioral, such as an instinct to mimic. Let's suppose here that the inherited feature is behavioral flexibility or "plasticity," and that it is enabled by a specific neural circuitry. Let's also assume that this neural circuitry is inheritable. This would mean that the capuchins and rabbits that inherited it were more likely on average to learn the new behaviors quickly and easily, and so more likely, again on average, to survive and reproduce. In each generation, plasticity would be present in more and more individuals. In each generation it would also be improved, so that—again, on average—the individuals that learned the new behaviors would do so more easily and more quickly.

This raises a question: Did natural selection select for greater plasticity and the neural circuitry that enabled the second sequence of events? Yes. But—and here's where organic selection comes in—that selection was prescribed by other selections. That first capuchin selected its nut-cracking technique, and that first rabbit selected its escape behavior. Organic selection then might be thought of as a version of natural selection, and in some ways a greatly improved version. It speeds evolution and makes it more efficient, producing an advantageous adaptation in fewer generations, with far less culling of individuals.

Some Animals (at Some Times) May Direct Their Own Evolution

Unlike natural selection, Baldwin's organic selection involves an actual, literal selection, with a selector (here the capuchin or rabbit) making a choice. The capuchin might have chosen other means to open those nuts, as the rabbit might have found other means to escape the predator, any of which might have had adaptive advantages. Those behaviors, too, would have been learned and imitated more easily by animals with greater plasticity, and they, too, would have led to an increase in number. "None of the above" is also a choice, and the capuchin and rabbit might have selected no new behavior at all. In that case, there'd be nothing to learn and imitate. Since plasticity would be of no advantage, it would not be selected for by natural selection, and subsequent generations would not on average show more individuals with it than without it. The capuchins' and rabbits' numbers would not increase and might well decrease. In making a choice (or making no choice), the capuchin and the rabbit would be setting a course for the evolution of their progeny.

In a "big picture'" sense—their conception of the nature of Nature—Baldwin's organic selection and neo-Lamarckian evolution overlapped. Neo-Lamarckian evolution posited that progress and free will were inherent in the evolution of all organisms, organic selection that they were important in certain cases.[17] While some naturalists labeled organic selection a compromise between the Darwinists and neo-Lamarckians, it wasn't really. Although it satisfied the neo-Lamarckian preference for a natural world that both progressed and allowed free will, the mechanism of evolution to which it appealed was natural selection. At its core, organic selection was thoroughly, determinedly Darwinian.[18]

In 1908 Baldwin was teaching at Johns Hopkins, where he became the focus of a scandal or would-be scandal and was compelled to resign.[19] The merits of the case against him are foggy, but clear is that the field of psychology in its nascent years was deprived of one of its most original voices. With his wife and two daughters Baldwin emigrated to Paris, where he continued to write and publish. But because he did so with a tarnished reputation and no academic affiliation, much of his new work was overlooked, and even his existing work began to receive less attention. In 1942 the hypothesis of organic selection was cited by Julian Huxley in his synthesis of natural selection and Mendelian inheritance,[20] but it was otherwise ignored. It began to receive attention again only in 1953, nearly sixty years after its introduction, when an article in the journal *Evolution* called it the "Baldwin effect," a name it has retained ever since.[21] In the 1980s researchers in human cognition and learning brought the Baldwin effect to bear on their findings, but evolutionary biologists and theorists continued to give it a wide berth. They thought it rare or nonexistent, and in either case unworthy of much attention. In the first decade of this century, though, scientists discovered evidence that bird species had learned behaviors that not only enabled them to greatly increase their range and their numbers, but also produced a change in an inherited feature.

Dark-eyed juncos are native to several mountain ranges in California, and some overwinter on the Southern California coast, a locale with a far-milder, Mediterranean climate. In the early 1980s, as spring began, some did not return to the mountains and made their permanent residence the campus of the University of California, San Diego. In 2004 researchers Pamela Yeh and Trevor Price found that the juncos' breeding season was longer than that of their highland relatives; where the mountain juncos had two broods, these at UCSD had time to produce as many as four. Thus the behavioral flexibility—or choice—of the juncos who remained near the coast year-round was rewarded with an increase in population.

That population underwent a physical change as well. In territorial fights and when courting mates, a male junco raises and fans open his tail, displaying white feathers; females have been shown to prefer males that have more white in the tail. Over several generations the tails of the male juncos showed significantly less white. Yeh and Price surmised that this change came about because the longer breeding season made parental investment more important than competition for mates. Females capable of producing many broods in one season became less inclined to mate with males with conspicuous tail feathers and more inclined to mate with males that had shown themselves to be capable providers for their first brood.[22] Consequently, more offspring were produced by unions involving males with less white in their tails, and over successive generations the tail feathers of the population as a whole showed less and less white.[23] Thus the choice of some juncos to make the Southern California coast their permanent residence led to a longer breeding season, which yielded a physical change in an inherited feature.

Many scientists today, perhaps most notably evolutionary theorist Daniel Dennett, regard the Baldwin effect as well established and important.

The Baldwin Effect and Music Appreciation

The Baldwin effect may have had a role in our own evolution—that is, yours and mine. Musicologist Piotr Podlipniak suspects that it might explain the development of *musicality*, the ability to recognize, learn, and perform music. Musicality has obvious adaptive advantages. It facilitates mother-infant bonds, enhances group consolidation, and increases sexual attractiveness; it has given us lullabies, school fight songs, and rock-star groupies.

But what is it? Consider that any given work of music is a product of the culture from which it arose. Reggae, for instance, grew from ska and other Jamaican variations on calypso and rhythm and blues. Musicality, though, transcends culture. It is what allows a reggae musician hearing a polka for the first time to recognize it not as mere sound, but as what Podlipniak terms "a syntactically organized structure based on pitch classes and rhythmic units measured in reference to a musical pulse."[24] That musicality is common to all humans implies that it is enabled by specific neural circuitry. If it is, says Podlipniak, then any piece of music upon its creation takes on an existence outside of and independent of its creator, and independent of us all. Then it doubles back and stimulates the development of a region in our brains and nervous systems. In creating music, inadvertently but quite effectively we may have directed our own evolution.

The Glove Turned Inside Out

What of the Baldwin effect and play? Curiously, not until 2001 did a group of ethologists make the connection that Groos and Baldwin had not. We've met them already: Spinka, Newberry, and Bekoff. In the same paper in which they proposed that an important adaptive advantage of play was training for the unexpected, they linked the Baldwin effect to play.

An animal engaged in functional behaviors—such as foraging and hunting—is set on specific tasks, and until those tasks are completed, it ignores or resists impediments and interruptions. It most certainly does not self-handicap. An animal at play, on the other hand, in yielding to and even welcoming impediments and interruptions, is far more likely than one engaged in functional behavior to produce idiosyncratic and innovative actions. Since the Baldwin effect is triggered by such actions, write Spinka, Newberry, and Bekoff, it is more likely to be set in motion by an animal at play than one doing anything else.

The Baldwin effect is likely to begin in play for another reason as well. Animals in a given population hunt, forage, and nurse their young in the same way. But animals that play often play differently. Moreover, innovations—new moves, new games—are introduced to the population by individual animals. Those 360-degree rotations observed by Petrů's team were being performed by only one de Brazza's monkey, and jumping in place with an object on its head was the recreational activity of only one Diana monkey. Innovations in the play of dogs and harbor seals, it seems, are likewise introduced by individuals.[25] This means that any population of playing animals is not only likely to produce more innovation than is a population of animals engaged in serious behavior; it is also more likely to produce more *kinds* of innovation.

Play, more than any other behavior, invites innovation, and over several generations any inherited features that enabled an animal to learn those innovations more quickly and easily would be selected for. Innovation produces the Baldwin effect, and the Baldwin effect accelerates natural selection. Thus the glove is turned inside out. Evolution gives rise to play, and in some cases play, or rather the animal that through play made that first innovation, sets a direction for its own evolution.

CHAPTER 10

Playing Animal

The peoples who inhabit the rugged shores of Scotland's Outer Hebrides, the west coast of Ireland, and the islands in the Faroe Archipelago share stories of the seal people, or *selkies*. One island in the Faroe Archipelago is Kalsoy, distinguished by a central spine running its nine-mile length that gives it the appearance of a half-submerged mountain ridge. Kalsoy is a singular place, and appropriately it has its own selkie story.

A young fisherman from the village of Mikladalur ventures to the rough west coast, where he sees seals emerging from the surf and moving about on the rock ledges. One begins to push against the rock, its dark and shiny coat splits, and a figure wriggles out, kicks the skin aside, and stands. It is a woman, naked and pale in the late-summer sun. The others do the same and spread their skins to dry. Soon they laugh and begin to sing and dance.

The fisherman is careful to stay hidden. He knows now that he is watching selkies. According to legend, the selkie's shed skin is what connects her to the sea. If he can keep it from her, she'll be unable to return and must become his wife. Moving quickly and staying out of the selkies' sight, he takes the first selkie's skin and waits. When the sun touches the horizon and the air has grown chill, the selkies end their revelries. That first selkie finds that her skin is missing and begins to cry. The others console her, but they know, as does she, that she cannot return with them. They slip slowly into their skins, and one by one they slide off the ledge into the surf, until she is left alone on the rocks, naked and shivering. The fisherman approaches her, awkward, fumbling. He holds her skin folded under his arm, but he offers her his coat, and she accepts it.

Years pass. The fisherman and his selkie wife manage a tolerable marriage. Each day he brings home the part of his catch that does not sell. She cooks, mends his clothes, sweeps their small house, and does all that might be expected of a fisherman's wife. But they talk little, do not laugh or sing, and he notices that she sometimes stops her chores and for whole minutes gazes at the sea. The fisherman keeps her skin locked in a chest and carries its key with him. But one morning as he departs, he is occupied with other thoughts— perhaps a poor catch the day before, rumors of a storm, or the cost of repairs to his boat. He is miles from shore before realizing he has forgotten the key. He returns to find the chest open and empty. So he crosses the island, knowing what he will find. There, on the rock ledge where he first saw her, are his wife's neatly folded clothes.

The fisherman eats alone. He sleeps alone. He cannot be consoled. Soon he comes to think of the sea as an enemy and grows half-mad. He begins to kill the seals that play on the ledges. He does not know which might be selkies, and perhaps he does not care. He cannot know that his selkie wife has taken a selkie husband and borne selkie children. One day, unwittingly, he kills her husband and two sons. The selkie wife vows revenge against the fisherman and lays a curse upon him and all the men of Mikladalur. One will die at sea, another by a fall from a cliff, then another and another, until so many will be dead that were they to stand and lock arms they would ring the island.

Many cultures have stories of humans taking animal form: Navajo skin-walkers turning into bears, Mesoamerican shamans taking the form of jaguars, and, perhaps most famously, Eastern European men and women becoming wolves. All such figures are what students of mythology call *therianthropes*, from the Greek terms *therion*, or "wild animal," and *anthropos*, or "human." Stories of metamorphoses in the other direction—animals becoming human—are probably as widespread, and in many of these, as with the selkie myths, the transformed animals become wives to humans. Korea, for instance, has traditions of snail wives, Japan and Russia have frog wives. As in stories of therianthropes, the transformation is inevitably incomplete; the wives' essence remains defiantly animal. It is one reason that few animal-wife stories end happily, and many end in violence. The violence that ends the story of the Kalsoy selkie wife, though, is of a shockingly greater magnitude. And so, perhaps, is the loss that incites it.

One imagines the fisherman at the moment he finds her clothes. He looks to the surf breaking against the ledges below and to the open sea beyond in the late-afternoon sun, its gray surface flecked with white. He knows that beneath is a world far vaster than his own. Unlike his island, it is boundless. So perhaps he allows himself to imagine a different ending. By luck or enchantment, he finds a sealskin, slips into it, and slides off the ledge into the surf. Somewhere in that great ocean he finds her. They reconcile. She shows him her world and its secrets, and in time he comes to know it as if it were his own.

But this is only a dream of what he can never know, never have. And this is what he loses, too.

A Child's Question

One suspects that humans have been imagining what it's like to be an animal, or what it's like for an animal to be an animal, for as long as there have been humans and animals. It seems a natural exercise of the imagination, especially the imagination of children. A child seeing a seal dive, a sparrow flit among branches, a goldfish nudging the glass of its bowl, is apt to ask what it feels like to be the seal, the sparrow, the goldfish. It's a fundamental question, and like many fundamental questions, it is often left unanswered. No one says it directly, but the response to the child comes in subtle changes of subject or silence. *You are not a seal or a sparrow or a goldfish. Think about something else.* But some children, it seems, choose *not* to think of something else. When scientists at a 1991 gathering of ethologists were asked why they had decided to study a certain animal, most answered that they wanted to know what it was like to *be* that animal.[1]

An Addition to Tinbergen

Recall that following the lead of Tinbergen, Burghardt and the Pellises contended that to fully understand an animal's play, you'd need to consider its adaptive advantage or advantages, how it developed throughout the animal's life, the physical features of the animal that made it possible, and its evolution over time. To these, Burghardt added that if you expected to fully understand an animal's play, you'd also need to know how the animal experienced it.[2] You'd need to answer the child's question. So in deference to the resolve of children who grew up to be ethologists, for the sake of argument, and in the spirit of open-minded (and playful) inquiry, let's explore the possibility that one *can* know what it's like to be an animal.

As Burghardt himself allowed, the knowledge would not come easily. It would be departing from, and indeed rebelling against, centuries-old ideas that any effort to understand an animal's inner experience would be pointless—for the simple reason that there was little there to understand. Several influential philosophers and theologians—St. Ambrose and St. Augustine in the fourth and fifth centuries and St. Thomas Aquinas in the thirteenth—held that animals differed from humans in that they lacked two attributes. One was an ability to reason. In this they were following Aristotle, who had defined human as "the rational animal" and by implication defined all others as irrational. The other attribute, rather more metaphysical and derived from Scripture, was a soul. They saw these attributes as conjoined: reason was what defined the soul.[3]

You and I might guess that a sheep running from a wolf it had never before seen was acting on an instinct, but these philosophers had no conception of instinct as such. They held that since a sheep does not possess reason, it cannot have deduced from prior experience that the wolf was a threat. Some explained the sheep's behavior—or rather tried to explain it—by inventing an attribute that looked like reason but was something else. It ran because it had perceived the wolf's lethal intentions through a kind of sixth sense they called *estimativa*.[4]

In the seventeenth century the French philosopher René Descartes proposed that "mind" was separate from the material universe, and the means by which humans were linked to the mind of God. Animals, who enjoyed no such connection, were automatons without souls, reason, or—minds. Not everyone agreed. Scottish philosopher David Hume countered that such ideas relied on a priori assumptions, were untethered from everyday experience, and paid little attention to what animals are actually like. In 1739 he dismissed these ideas out of hand, declaring, "No truth appears to me more evident than that beasts are endowed with thought and reason as well as man." He continued, "The arguments are in this case so obvious, that"—here a rhetorical missile that seems to have been aimed squarely at Descartes—"they never escape the most stupid and ignorant."[5]

By the late nineteenth century Hume's commonsensical understanding was widespread. Edward Thompson, writing in 1851, was convinced that animals possessed what he called "an inward sense, analogous to the faculty of the soul."[6] So many natural philosophers held that animals possessed minds and inner experiences that in 1871 Darwin declared the question settled: "It has, I think, now been shewn that man and the higher animals . . . have the same senses, intuitions and sensations—similar passions, affections, and emotions, even the more complex ones; they feel wonder and curiosity; they possess the same faculties of imitation, attention, memory, imagination, and reason, though in very different degrees."[7] In his 1872 work *The Expression of the Emotions in Man and Animals,* he went further, arguing at length that the behavior of many animals could be understood *only* by acknowledging their thoughts and feelings.

A book that traced their evolution would make for a logical sequel to *Descent* and *Emotions,* but by the late 1870s Darwin was directing his interests toward what would be his final book, *The Formation of Vegetable Mould through the Action of Worms.* He entrusted his ideas and notes on the evolution of animal minds to another—his colleague and protégé Canadian-English naturalist George Romanes.

George Romanes and Persons Bearing Names Unknown to Fame

Like Darwin, Romanes held that there existed "a psychological, no less than a physiological, continuity extending throughout the length and breadth of the animal kingdom,"[8] and he undertook work that might demonstrate that continuity. He began by collecting reports from far-flung correspondents, planning to accept only those from well-known and respected naturalists or persons that naturalists recognized as credible. But he soon found that the subject of animal minds was difficult to treat with such rigor. Empirical tests were rare, but intriguing anecdotes abundant. And of those anecdotes the most intriguing were from layfolk—or as he put it, "persons bearing names more or less unknown to fame."[9] Romanes suspected that some accounts had real value and worried that in dismissing them out of hand he might overlook something important. So he adopted an editorial standard. Of the most intriguing accounts he would accept only those that described behavior "particularly marked and unmistakable," that gave all appearance of being accurate and unbiased, and that could be corroborated by similar accounts.

The result, his 1882 work *Animal Intelligence*, overflows with anecdotes: a baboon taking revenge, an elephant concealing a theft, birds engaging in piracy, and swans demonstrating conjugal fidelity. One spellbinding account tumbles after another, and the reader comes away dazzled, all the more because most are credible, some excerpted from the writings of respected naturalists of Romanes's day, others verified by ethologists in later periods. Still, many of his contemporaries criticized the book, not because they found its accounts dubious, but because they thought the method by which he gathered them was myopic. He seemed not to have considered the distinct possibility that his sources had reported only the behaviors that seemed to display intelligence and had ignored all others. Moreover, he said nothing about the frequency of the more remarkable instances and so left open the possibility that they were rare or even unique.

The inadequacies of Romanes's work—anecdotes presented as evidence, and a general absence of rigor—were the banes of many scientific fields. But animal behavior was also much vexed by another quandary, one particular to itself. In the 1890s many naturalists grew concerned that some of their colleagues—Romanes among them—were attributing human qualities to animals and thus indulging a mindset and practice called *anthropomorphism*. The term was decidedly pejorative, carrying an implicit critique of the attributor as unscientific.

The Clever Hans Affair

The dangers of attributing human qualities to animals and the necessity for rigor in studies of animal behavior would soon be made dramatically evident both to scientists and—since the concerns were not merely academic—to the public at large. Some members of that public were particularly interested in the performances of a Russian trotting horse named Clever Hans. In the early 1890s Hans performed at exhibitions in Berlin. With hoof taps or nods of his head, he would respond to questions from his trainer and persons in the audience. He seemed able to calculate mathematical problems, to understand German, and to read musical notation. Some members of the small community of scientists interested in equine intelligence suspected that all was slight of hand and hoof. They were not alone. The public interest in Clever Hans's performances was so great that the German board of education appointed a commission—made up of a psychologist, two schoolteachers, a circus manager, two zoologists, and a horse trainer—to investigate them. The commission observed Hans's performance, found no subterfuge, and concluded that the horse's displays of intelligence were genuine. Their testimony persuaded many. But not all.

Oskar Pfungst, a student at the Psychological Institute at the University of Berlin, remained skeptical. With the trainer's permission and cooperation, Pfungst undertook his own series of tests. He found that Hans could answer questions correctly only if the questioners knew the answer, and only if the trainer or questioners were visible to Hans. Pfungst also noted that as the questioners and trainer awaited Hans's response, they stood rather stiffly, and when he arrived at the correct answer, they unconsciously relaxed their posture. The questioners and trainer, Pfungst realized, were unwittingly giving the horse cues, to which the horse was responding. The episode became a cautionary tale to those who would credit anecdotes of animal behavior uncritically, and Pfungst's observational method, in its rigor and attention to detail, would come to be regarded as a model of experimental design.

The Clever Hans affair made clear that the study of animal behavior needed a guiding principle. In 1894 psychologist C. Lloyd Morgan supplied one: "In no case may we interpret an action as the outcome of the exercise of a higher psychical faculty if it can be interpreted as the outcome of the exercise of one which stands lower in the psychological scale."[10] Given a choice of explanations for an animal's behavior, Morgan said, the simpler was to be preferred. He had applied Occam's razor to the unkempt hypotheses of the field; in time his principle came into widespread use, becoming known as Morgan's canon.

Methodological Behaviorism

Anthropomorphic thinking was not easily quelled, however, and continued to vex the nascent field of animal behavior. Groos, writing in 1898, criticizes one of his sources as "marred by the attempt to humanize the actions of animals."[11] As a corrective, some scientists limited their scope to stimuli and behavior that could be observed, a practice called *methodological behaviorism.* It had virtues. It simplified and clarified research methods and enabled results to be quantified. It also made it easier for researchers to exclude their own thoughts and feelings from their findings. For the first half of the twentieth century, the practice guided much research in animal behavior. But in the second half of that century, some researchers became increasingly aware that it was not quite up to the task. Animal behavior had a richness and complexity that methodological behaviorism could not explain.

In a 1976 book entitled *The Question of Animal Awareness,* zoologist Donald Griffin called upon mounting evidence from neurological and behavioral studies that showed animals are self-aware and conscious and admonished a scientific culture unwilling to consider that evidence. Scientists who denied animal consciousness, he said, were failing to recognize certain qualities in animals for no reason other than that humans possess the same. Their stance, he believed, was an impediment to scientific progress, every bit as blinkered and unscientific as attributing human qualities to animals. He called it *mentophobia*. Had the rebuke come from a less eminent scientist, it might have been ignored. But Griffin had discovered echolocation in bats. For that among other reasons, he was renowned and respected. His book persuaded many that a rethinking was overdue. Before long, it inspired a new field, *cognitive ethology*, the study of the mental states of animals based on their interactions with their environment.

Still, methodological behaviorism was not abandoned entirely. As recently as the late twentieth century, a much-respected text-book counseled, "One is well advised to study the behaviour rather than attempting to get at any underlying emotion."[12] Resistance to acknowledging animals' inner lives, and attempts to study those lives, persists. The legacy is rooted in the philosophy and theology—and so the cultural traditions—of Western Europe. But other cultural traditions hold animals and their inner lives in higher regard, and recognize no appreciable distinction between their minds and our own. Buddhism, for instance, maintains that a human soul may be reborn in the body of a nonhuman animal. That the belief underlies much of Japanese culture may have made it easier for one scientist living within that culture—Japanese ecologist and anthropologist Kinji Imanishi—to recognize that primates display cultural differences within species.[13] When Imanishi first presented his discovery, some primatologists thought it suspect. The reason, so said primatologist Frans de Waal, was that they were still under the thrall of the Western European traditions, a sort of thinking he termed *anthropodenial*.

Burghardt's Middle Way

As ideas that the behavior of animals and humans might share roots—ideas long thought anathema—were reassessed, so, too, was the value of anecdotal evidence. By the 1990s Burghardt, long a critic of such evidence, was coming to believe it useful, if not indispensable, to the study of animal play. There were several reasons. First, if used judiciously, anecdotes could help to clarify and illuminate the qualitative data that might otherwise be lost among the charts and bar graphs. Second, while a single anecdote of, say, deception in primates might be of dubious value, collections of anecdotes of the behavior could be subject to rigorous quantitative analysis. (This practice has already proved its worth. Biologist and comparative psychologist Louis Lefebvre was interested in innovative behavior of birds, most reports of which were anecdotal. He scanned seventy-five years of bird behavior and ecology journals for words such as *novel* or *unusual* and found twenty-three hundred examples of innovative behavior from hundreds of species.) [14] The third reason was that while a single anecdote might be of little value in and of itself, it might lead scientists to look for behavior they otherwise wouldn't have. And fourth, Burghardt thought anecdotes useful because they were often the only data available.

More recently, Burghardt has suggested that ethologists can't help but imagine that emotions like those that govern their own behavior also govern the behavior of an animal, and that for anyone wishing to understand animal minds, anthropomorphism may be the default starting position. Burghardt says that is not necessarily a bad thing. When used judiciously and in moderation, such an approach might provide real insights. He suggests that ethologists, using a technique he calls *critical anthropomorphism*, might examine their intuitions and compare them to those of other ethologists. Shared intuitions would less likely be projections of any individual ethologist, and more likely to represent the animal's actual experience. Those intuitions were worth taking seriously and might be used to develop hypotheses and to direct further inquiry.

Eiseley and the Crow

One intuition worth taking seriously was that of anthropologist and author Loren Eiseley. Of his many accounts of encounters with the natural world, one is particularly relevant to our concerns here. Eiseley was walking along a country road in a dense fog when suddenly a large crow flew a few feet over his head and let out a shrieking caw. The sound was so piercing, Eiseley thought, that it could only have been a cry of outright terror. The crow, disoriented in the fog, must have believed it had encountered a man walking in midair, and such a thing was, in Eiseley's haunting phrase, "the most profound evil a crow mind could conceive."[15]

We might suppose that a moment of insight into an animal mind should be like this—startling, even shocking. Eiseley's intuition *seems* right. But was it? Or was he imagining only what he wanted to imagine? We can't know. But by taking the approach Burghardt recommends—in this case by comparing others' experiences with shrieking, disoriented crows—we might build a case that Eiseley was onto something.

The "Surround Worlds" of Jakob von Uexküll

In efforts to determine the nature of an animal's experience, critical anthropomorphism—that is, inquiry sprung from hypotheses derived from intuition—would be a logical place to start. But no more. In the 1920s, one researcher had proposed a way to build upon those intuitions, extend them, and test them. He was an Estonian-born German named Jakob von Uexküll, and his scholarly interests might be called philosophical biology.

Von Uexküll argued that an animal, any animal, perceives the world through signs it finds of interest and is insensible or oblivious of all others. A goshawk might note a slight movement of grass in a meadow that may indicate the presence of a field mouse, but pay no attention to newly flowering red clover nearby. A bee would notice the clover, but perhaps not the movement in the grass. Every animal may be said to live within a bubble of its own perceptions. Von Uexküll called such a bubble an *umwelt* ('ŏom͵velt), literally "environment," or more literally "surround world." The notion gives rise to a dizzying vision of animal life on earth as bubbles of wildly varying sizes. An amoeba's umwelt might be only slightly larger than its body, while an arctic tern flying over open ocean and navigating by distant stars might have an umwelt hundreds of light years across. Since umwelten move with their subject and often interpenetrate, at any moment countless umwelten are sliding over, under, and through one another, and each has an animal at its center.

Von Uexküll developed the idea over much of his career, and later formulations with two layers of umwelten—*receptors* that received information and *effectors* that acted upon it—were complex enough that their full explication required more than seventy specialized terms. The idea influenced the study of signs and symbols called semiotics. It also anticipated concepts of information processing now widely employed in the fields of robotics and artificial intelligence.

Anthropomorphism by Omission

But umwelten, while useful, might also mislead. If we imagine our own umwelt bubble overlapping some part of an animal's, we see only our bubble and the slice of our bubble that overlaps the animal's. It would be easy to imagine that this overlap represents the animal's entire experience, but we'd be missing the potentially vast part of the animal's umwelt that lies outside ours, the part that English poet William Blake called "an immense world of delight, clos'd by your senses five." Unless we acknowledge *all* of the animal's experience, that immense world of delight, we would fall into another sort of anthropomorphism—not by imposing attributes that aren't there, but by failing to recognize attributes that are. We would be committing what Burghardt calls *anthropomorphism by omission*. This was exactly the sort of error the audiences of Clever Hans had made. They gauged Hans's intelligence with their own indices of intelligence: numerical and language skills. Hans was proficient in neither. He was, though, quite proficient in recognizing *body* language, and that sort of intelligence was clearly lacking in the trainer and audiences.

The brains of most animals are, on balance, simpler than ours. Yet it may not follow that their sensory experience is poorer. Owing to the organs with which they mediate that experience, it may be far richer. A falcon can see a mouse from a distance of up to a mile. Some animals, such as Panksepp's rats, have ears that detect ultrasonic frequencies. The "nose" of the star-nosed mole is an organ of touch; it has five times the number of touch receptors that you have in one of your hands. Elephants, through their feet and trunk, can detect reverberations ten miles away, and by comparing subtle differences in what each foot feels, can triangulate their source. And then there's smell. By comparison with many animals, we are nearly anosmatic. Our noses have perhaps 6 million olfactory receptors, but a dog's nose may have as many as 300 million. A silkworm moth can detect pheromones from seven miles.

Although a falcon's sight, a mole's touch, and a moth's smell are far more acute than their counterparts among our "senses five," they are essentially the same senses. Some animals are endowed with a sensory repertoire far wider. Pit vipers have membranes, one below each nostril, with which they can see infrared light. A shark can hunt in murky water because a network of pores in its head are filled with Lorenzini jelly, a biological conductor of electricity that enables the shark to sense small differences between the electrical charge of the prey and the water around it. Honeybees can detect the earth's magnetic field, by means scientists don't fully understand, but perhaps with a magnetic mineral that lines cells in their abdomens. When the bees fly, they accumulate a small positive charge. Because hairs on their legs respond to the attraction between this charge and the negative charge of a flower, the bees can use the hairs to guide them to a bloom. Once a bee visits, the flower's charge changes. And that, if you've ever wondered, is how a following bee knows a flower's store of nectar has been depleted and can take a pass on it.

Such senses may be simulated by virtual reality systems. One allows humans to navigate a virtual environment using bat echo-location sounds.[16] Such sounds are ultrasonic, and the system does not enable subjects to hear them directly, but rather converts them to lower frequencies audible to a human ear. Panksepp's repurposed bat detector worked much the same way. Likewise, night-vision goggles don't enable us to see infrared light as a pit viper sees it, but to see it converted to a range visible to us.[17] These technologies, impressive as they are, don't convey to us an animal's experience so much as they translate that experience into one we can comprehend. In those translations, it's safe to say, much is lost.

"The Eye of Man Hath Not Heard, the Ear of Man Hath Not Seen..."

It's one thing to know what it feels like to be an animal. It's quite another to describe that feeling. In Shakespeare's *Midsummer Night's Dream*, Nick Bottom spends most of the play with his head transformed into that of a donkey. When he is returned to his fully human state, he exclaims, "The eye of man hath not heard, the ear of man hath not seen, man's hand is not able to taste, his tongue to conceive, nor his heart to report what my dream was." Bottom's synesthesia—eyes that hear, ears that see, and hands that taste—might be an aftereffect of the experience, his senses still tangled and confused. It might also suggest that the experience of being an animal exceeds the descriptive powers of language.

Nonetheless, a great many works of literature attempt to represent that experience. Among the better known are Leo

Tolstoy's 1886 "Kholstomer: The Story of a Horse" and Virginia Woolf's *Flush: A Biography*, the life (as imagined by Woolf) of Elizabeth Barrett Browning's cocker spaniel. More recent works include an edited collection, *Unleashed: Poems by Writers' Dogs*, which features a composition by an Irish setter, its single succinct line: "Leaves—I thought they were birds." Not all who engage in such exercises are poets, but many who do are endowed with a poet's imagination. Margaret Floy Washburn was an American psychologist, and in 1921 the second woman to serve as president of the American Psychological Association. In her book *The Animal Mind: A Textbook of Comparative Psychology*, she imagines what it's like to be an amoeba, a creature whose mind, if it can be said to have one, is unencumbered by memory: "The Amoeba's conscious experience may be rather a series of 'flashes' than a steady stream. . . . Each moment of consciousness is as if there were no world beyond, before, or after it."[18] A scientist who has made more recent investigations into animal minds—particularly that of *Canis familiaris*—is Alexandra Horowitz, whose delightful *Inside of a Dog: What Dogs See, Smell, and Know* describes the canine experience in all its strangeness and wonder.

If we allow an animal sentience and emotions, recognize its entire umwelt, employ assistive technologies where necessary, and temper our imagination with critical anthropomorphism, we might gain some insight into its experience. But is it enough insight? Can we really know what it's like to be an animal? Philosopher Thomas Nagel said no. In fact, that summarizes his widely read and much-cited 1974 article entitled "What Is It Like to Be a Bat?" Nagel held that the experience of being a bat is the sum of input from the bat's senses, some of which, like echolocation, are quite unlike our own and are interpreted by a mind also quite unlike our own. That experience, Nagel said, is separated from ours by a vast and unbridgeable gulf. Summoning all the resources of your imagination, you might hypothesize what it might be like for *you* to be a bat, but you cannot imagine what it is like for a *bat* to be a bat.

Still, one might try. To understand an animal's experience from within, one would need to live as much as possible *as* that animal. Anyone doing so, or attempting to do so, would be endowed with curiosity, a rare freedom of spirit, and an unusual sort of courage.[19] Or perhaps just be a bit eccentric.

The Man Who Was a Goat

Thomas Thwaites is a graduate of London's Royal College of Art who describes himself as "a designer (of a more speculative sort), interested in technology, science, [and] futures research." One day several years ago he was walking his niece's Irish terrier. He was going through a period of existential angst, and as he watched the dog sniffing grass and feeling the wind, he realized that it was free from all the worries that plagued human existence.* He found himself envious.

Many of us have had similar thoughts. But Thwaites, true to his venturesome character, acted on them. He devised an experiment in which by living as an animal—that is, getting inside the animal's experience as much as possible—he might free himself

---

* Thwaites's thoughts on the advantages of an uncomplicated lifestyle has literary precedent. In May Kendall's droll (1885) "Lay of the Trilobite," the poem's human speaker, in what may be his own imagining, is persuaded by a fossil trilobite that its existence might have been happier than his.

from existential anxieties, at least momentarily. First, he needed to decide upon an animal. An early possibility was an elephant, but he visited a shaman in Copenhagen who counseled that an Englishman has no business trying to be an animal whose natural habitat is so unlike his own.

"She told me I should try and become a goat," Thwaites said. "I think she was quite right."[20]

Thwaites approached the challenge systematically. To learn to perceive the world as a goat might, he began reading philosophy. Upon finishing Martin Heidegger's *Being and Time*, Thwaites decided that he would need to abandon cognitive habits he didn't realize he had. He'd need to see a chair without thinking of sitting, see a word without reading it, and see a goat not as a goat, but as another person. To learn to move like a goat, he solicited a maker of prosthetic limbs to fashion extensions for his arms and legs. Since eating occupies much of a goat's waking existence, it's a particularly important aspect of goatness. But the diet of a goat presented Thwaites a particular challenge—he could not digest grass. So he allowed himself a compromise. During the day he would chew grass, spit it into a container, and save it for night, when he would make it digestible by stewing it in a pressure cooker.

The experiment began, with the permission of an obliging and good-humored shepherd, among a herd of goats on a mountain in Switzerland. Thwaites wore the goat prostheses on his hands and feet, a Gore-Tex jacket, and a bicycle helmet. On the first morning, when the goats were being herded to pasture farther down the mountainside, he struggled to keep pace. The effort was exhausting and, since the slope was steep and strewn with boulders, at times terrifying. But after a few days he became acclimated to the terrain, and to his ruminant companions. Of a certain goat he said, "I wander after her when she moves to another patch of grass." The amicability was returned. "And likewise when I move off, she's not far behind."[21]

A Badger, an Otter, an Urban Fox, a Red Deer, and a Common Swift

Another recent venture into the experience of animalness was undertaken by British veterinarian, author, and sometimes dweller of badger setts Charles Foster. His motivation was rather the reverse of Thwaites's. Whereas Thwaites desired to escape the human experience, Foster longed to enter the animal's. In his enchanting book *Being a Beast*, Foster recounts a series of attempts to apprehend the experience of an otter, an urban fox, a red deer, a common swift, and a badger. Like Thwaites, Foster approached the challenge with logic. He reasoned that we are a social species, and like most social species, we have a well-developed theory of mind and thus are unusually well equipped to imagine the worldview of another, even when that other is nonhuman. He further reasoned that much of what any animal experiences is not so much the sum of input from its senses, but what part of that input it pays attention to. And he could, through study, learn what part that might be.

A badger, Foster knew, lives in a world of scent. Foster allowed that his own sense of smell might be dull by comparison, but worked to improve it by making a practice of sniffing his children's soiled laundry until he could tell who wore what. Sometime later, as he crawled through tall grass of the Welsh countryside, his practice was rewarded. He detected "the citrusy piss of the voles in their runs within the grass; the distantly maritime tang of a slug trail, like a winter rock pool; the crushed laurel of a frog; the dustiness of a toad; the sharp musk of a weasel; the blunter musk of an otter."[22] On cold, dry mornings, he noted that smells hug the ground, staying near their sources, and as the day warms, they rise and swell. He surmised that for a badger scents have shape. A tree on a warm day is "the helical shape of the scent vortex that pulls dust up into the canopy"; on a cold one, it's "a low hump of tart lichen with an indistinct chimney."[23] He ate what a badger eats, savoring so many earthworms that he soon became an annelid connoisseur, noting that some are musty, tasting of decay and splintered wood, and others, when chewed, have a long mineral finish.

Since badgers are largely nocturnal, Foster undertook most of his olfactory and gustatory activities at night and spent his badger days accompanied by his intrepid and obliging eight-year-old son, sleeping in a trench dug in a hillside as a makeshift badger sett. Foster knew full well that the undertaking was quixotic, a bit daft, and perhaps destined to fail. His recounting is not without humor. One afternoon his friend Burt, whose backhoe dug the trench, brought them fish pies. Foster accepted the pies, assuring his son, friend Burt, and the imagined and perhaps admonitory reader that since no sane badger would refuse the pies, he had in no way compromised the experiment's authenticity. Nonetheless, Burt sensed some lingering guilt; to relieve it and to give Foster a chance to regain any lost badger bona fides, Burt offered to set his dogs on Foster or try to hit him with his truck.

As human behaviors, the endeavors of Thwaites and Foster were unusual. How might we categorize them? Our first thought might be to think of them as exploration and investigation. But that isn't quite right. Exploration and investigation answer immediate needs of survival or long-term needs of reproduction, and they have well-defined end points. Thwaites and Foster were answering no immediate or long-term need, and while Thwaites's venture into "goatness" lasted six consecutive days, and Foster's venture into various "animalnesses" occurred at intervals over several years, neither had a predetermined end. Their behavior did, however, meet Burghardt's definition of play—it was nonfunctional, voluntary, characterized by repeated but varied movements, and undertaken when their performer was well-fed, safe, and healthy. Moreover, since Thwaites and Foster put themselves in positions that were disadvantageous and even dangerous, their endeavors required self-handicapping, the feature Spinka and company believe essential to much play. We might further conjecture that the reason they engaged in their behavior was at its core the same reason that puppies leap into snow and patas monkeys practice parkour: because to do so is interesting and exciting. We might venture that the reason natural selection selected their behavior was the same reason it selected the behavior of the puppies and the monkeys—to allow them to learn something new.

What did Thwaites and Foster learn? Thwaites found, he said, that "the way we perceive the world is both fixed and flexible."[24] He still perceived a chair as an object for sitting on. But when he strapped on four legs, he couldn't use his hands, so his mouth became his "interface with the world."[25] Foster had no single epiphany but did experience several small ones, one of which occurred on an evening while he was pawing through garbage in an East London alleyway, a necessary part of the experience of being an urban fox. He saw the glimmering lights from televisions in nearly every house. To Foster—or rather Foster as a fox—the people in those houses were oblivious of their place—living surely, but living nowhere in particular. The foxes in the neighborhood, he thought, were its true residents. They knew "there's a mouse nest under the porch at number 17A and bumblebees by the cedar decking at number 29B."[26] When his efforts failed, Foster also learned humility. To understand a swift, he cultivated an awareness of air currents and studied flight paths. But that knowledge gave a mere hint at what being one must be like. In their migration, the birds fly nonstop for as many as three hundred days and nights over open ocean—a feat, and so an experience—quite beyond his ability even to imagine. As for being a swift, he wrote, "I might as well try to be God."[27]

Interspecies Play

Elisabetta Palagi, a professor at the University of Pisa, was particularly curious about play between species. She and her research team studied videos of dogs and horses in play with one another, teasing, nipping, pulling away for a moment and then returning to tease again. Her team found that in such interactions both animals self-handicapped, shaking heads and rolling on their backs. Remarkably, the self-handicapping was bespoke. Each was holding back, moderating its own advantages to accommodate the other—the dogs nipping not biting, and the horses, with the advantage of height, lowering their heads and shoulders and sometimes lying down.

Many animals show they are playing with a facial expression called the *relaxed open-mouth display*. The mouth is opened only slightly, and in animals with canine teeth not so wide as to show those teeth. The behavior has long been observed in a range of mammals, among them primates, otters, American black bears, and horses. But until Palagi's work with dogs and horses, the behavior had been studied only among animals of the same species. She and her team found something new. Here were members of two species—one from an evolutionary lineage of predators, the other of prey—playing together. As surprising was that each used the relaxed open-mouth display to signal its intentions. One would display "I am still playing," and the other would, in a behavior called rapid facial mimicry, imitate it immediately, as if to say, "I know." Palagi and her team concluded, "Despite the difference in size, the phylogenetic distance, and differences in the behavioural repertoire," dogs and horses play.[28]

Play offers a means to bridge the gulf between species. In play, each animal becomes aware of the other's abilities. To self-handicap appropriately—that is, in response to a play partner's vulnerabilities—each must recognize those vulnerabilities. To predict a play partner's moves and to act on those predictions, each must practice a theory of mind; each must understand the other's point of view and perhaps, for brief moments, even adopt it. When animals play, all these skills are brought to the fore easily, naturally, and without conscious effort. When two animals play together, each learns something of what it feels like to be the other. When dogs and horses play, the dog learns what it feels like to be a horse, and the horse learns what it feels like to be a dog.

Play, Life, and Everything

By the last decade of the nineteenth century the neo-Lamarckian, metaphysical concerns with the nature of Nature were regarded as Darwin had regarded questions about a deity—that is, as beyond the reach of scientific inquiry.* In 1910 philosopher and educational theorist John Dewey observed that the concerns were nearing the end of their useful life and predicted how they'd reach it: "The displacing of this wholesale type of philosophy will doubtless not arrive by sheer logical disproof, but rather by growing recognition of its futility."[1] Dewey's conjecture has proved on the mark. Most scientists nowadays stay near the terra firma of their specialty and leave the large what-it-all-means questions to theology, to theology's more empirically minded cousin cosmology, and to looping, late-night bull sessions that my students assure me—and I am gratified to know—still

---

* "I feel most deeply that the whole subject is too profound for the human intellect. A dog might as well speculate on the mind of Newton," Darwin wrote. Letter to Asa Gray, May 22, 1860, Darwin Correspondence Project.

occur. I tend toward an early-to-bed pragmatism, but I find myself hoping that there is a nature of Nature, and that we might one day understand it.

Play gives us a hint not of the nature of *all* Nature, but perhaps much of it. Although we can't say with certainty what play is, we can say what it is *like*. It is like natural selection. Both play and natural selection are purposeless, ongoing, open-ended, and at any given stage provisional. In the short term, both are wasteful and profligate even to the point of extravagance. Both experiment, producing many outcomes that are useless or detrimental, but producing a few that in time prove beneficial and necessary. Both bring order from disorder, establishing basic patterns that are reshaped and reused, but seldom discarded completely. Both create beauty. Both hold forces of competition and cooperation in a dynamic equilibrium. Both employ deception. And both can operate without a material form.

To many biologists, the best definition of life is that which evolves by natural selection. Since natural selection shares so many features with play, we may with some justification maintain that life, in a most fundamental sense, is playful. The resemblance is not perfect. In one significant way natural selection and play are dissimilar: natural selection occurs over vast time spans and so is largely invisible, whereas play is quite visible. But since the characteristics of the prolonged process of natural selection are evident in an instance of play—as if millennia were squeezed into minutes—the dissimilarity offers insight. A new thought presents itself. Perhaps the reason we are so beguiled by watching animals at play is because when we do, we are seeing natural selection—and so life itself—distilled to its essence.

*   *   *

Ethologists have proposed a number of definitions of play that have proven useful, if not essential, to their study. Yet in privileging scientific rigor, those definitions may seem unduly materialist—like describing a leaf by listing its constituent chemicals or defining a rainbow as the refraction of sunlight by atmospheric water droplets. We might wonder whether the definitions are missing something. Play has about it a quality of the mystical and transcendent, a quality halfway to the spiritual.

The Sacred and the Profane is a treatise on religious experience by Mircea Eliade, one of the most influential scholars of religion of the twentieth century. Eliade hypothesized that humans of premodern societies saw much of the world as a chaos beyond their control, at times terrifyingly so. Yet places within that world—a certain tree, a certain mountain—and moments within that world—the time of harvest, or a new moon—could be loci of stability, offering access to a deeper reality that was ordered. At these places and moments, premodern humans performed rituals that provided respite from chaos of mundane space and time and gave them access to that deeper reality.

Johan Huizinga's *Homo Ludens* might be *The Sacred and the Profane*'s companion piece, or its slantwise sequel. Huizinga observed that the space in which play is performed, whether board or playing field, is circumscribed by what he called a "magic circle." Within that circle rules are established and obeyed, players who disregard those rules are penalized or ejected from the play, and for the duration of play players ignore everything outside the circle. This experience, and its partitioning of space and time, is well-known to followers of what's been called "America's pastime."

A baseball field is an ordinary piece of real estate bounded by property lines—a finite space. Yet players and spectators may experience that same space as unbounded—its baselines stretching from home plate outward and, in theory, to infinity. The time allotted a game is likewise bounded in a mundane, workaday sense—beginning at a certain hour and ending at another. Yet players and spectators may experience that same time as stretched or suspended, and see their experience reflected and lent weight in the game itself. Baseball uses no clock, and since it allows for unlimited extra innings, the final out may never come, and any given game might be played into eternity.

The cosmologies of Eliade and Huizinga are analogous. Games, like religious rituals, bring order to an otherwise chaotic world, and games, like religious rituals, allow their enactors an experience that is, or may be, transcendent. That animals forgo the needs of their bodies to play—playing when they are tired or hungry—suggests that the experience for them, too, is transcendent.

What then is play? Since play needs a body yet transcends the body, it is at the least a thing that harbors a contradiction—or several. Play is at once cooperative and competitive. It's by turns destructive and constructive. It enables its participants to disregard societal conventions, yet it assigns them fixed roles and imposes strict rules. All this might seem to make any effort to define play, much less understand it, futile. Perhaps it is. But that play is made of contradictions is not necessarily cause for discouragement.* The nature of religious experience is likewise

---

* To quote physicist Niels Bohr on a similarly ambiguous subject, "How wonderful that we have met with paradox. Now we have some hope of making progress." Ponomarev and Kurchatov, *Quantum Dice*, 75.

paradoxical and so ineffable. The koans of Zen Buddhism—the best known of which may be "When both hands are clapped, a sound is produced; listen to the sound of one hand clapping"—are intended to show that enlightenment is beyond the reach of language and reason. The only way to understand a religious experience is to have one.

It may be likewise with play. Perhaps the thing we are trying to understand is also the means to understand it. We understand play best by playing.

Tossing a ball with, say, a border collie, for as long as the tossing and retrieving lasts, we put aside worries about appointments and responsibilities. The border collie likewise ignores its usual canine concerns—perhaps incipient thirst, or a nearby squirrel. Tossing and retrieving, retrieving and tossing, our identities as human and canine become insignificant, and whatever Linnaean membrane might separate us melts away. For a long moment, a moment that seems timeless and eternal, we are not human and canine in a grassy area in a city park; we are only two players in a game.

Acknowledgments

Thanks to Gordon Burghardt, Robert Fagen, and Sergio Pellis for sharing their insights, and to the many other researchers whose works I drew upon. To innumerable non-human animals with whom I've played, most especially Lily, Sam, Houdini, and Lazlo. To the Cheney Agency for taking on the project, to my agent Allison Devereux for representing it so well, to editor Colin Harrison for overseeing its development, to Laura Wise for expertly managing its production, and to editor Emily Polson for her unfailing attention to matters large and small. A special thanks to both Allison and Emily, who gently nudged me to find the book's thesis. Much of what is good in the book is owed to them. Lastly, to Julie Hines. Of human animals, one of the most playful I've been fortunate enough to know.

Notes

INTRODUCTION

1 Burghardt, *Genesis of Animal Play*, 7.
2 "Science of the Brain," 252.
3 Burghardt, *Genesis of Animal Play*, 7.
4 Fagen, *Animal Play Behavior*, 494.
5 Wilson, *Sociobiology*, 84–86.
6 Darwin, *Origin of Species*, 414.

CHAPTER 1 Ball-Bouncing Octopuses: What Is Play?

1 Mather, "Octopuses Are Smart Suckers!?"
2 Godfrey-Smith, *Other Minds*, 43.
3 Boal et al., "Experimental Evidence for Spatial Learning." In the 1970s some researchers studying octopus behavior used electric shocks and surgically removed nerves and parts of the brain with no anesthetic. More recently, in part owing to a greater awareness of animal consciousness generally and octopus awareness specifically, researchers have employed less intrusive and more humane methods. In 2012 an international group of neuroscientists reviewed research on the neurobiological basis of consciousness in humans and nonhumans. One product of their work was the Cambridge Declaration on Consciousness, which reads in part, "The weight of evidence indicates that humans are not unique in possessing the neurological substrates that generate consciousness. Non-human animals, including all mammals and birds, and many other creatures, *including octopuses* [my italics], also possess these neurological substrates."

4 Ayala and Rzhetsky, "Origin of the Metazoan Phyla."
5 Mather and Anderson, "Exploration, Play and Habituation."
6 Borrell, "Are Octopuses Smart?"
7 In their article "On the Functions of Play and Its Role in Behavioral Development," Martin and Caro define play as "all solitary activity performed postnatally that appears to an observer to have no obvious immediate benefits to the player, in which motor patterns resembling those used in serious functional contests may be used in modified form."
8 Bekoff and Byers, "Critical Reanalysis."
9 Immelmann and Beer, *Dictionary of Ethology*, 223.
10 Mason, "Stereotypies."
11 "Science of the Brain," 269.
12 Tzar and Scigliano, "Through the Eye of an Octopus."
13 Ibid.
14 Kuba et al., "Looking at Play in *Octopus vulgaris*."
15 Ibid.

CHAPTER 2 The Kalahari Meerkat Project: The Hypotheses of Play

1 Byers, "Terrain Preferences."
2 Hausfater, "Predatory Behavior of Yellow Baboons."
3 Harcourt, "Survivorship Costs of Play."
4 Thompson, *Passions of Animals*, 61.
5 Schiller, *On the Aesthetic Education of Man*, 207.
6 Darwin, *Descent of Man*, pt. 2, 54.
7 Ibid., 7.
8 Groos, *Play of Animals*, xx.
9 In 1904 American psychologist G. Stanley Hall posited the *recapitulation theory* of play. It ran counter to Groos's hypothesis that play was preparation for the future, suggesting rather that much play was a vestige of the past. For instance, "The power to throw with accuracy and speed was once pivotal for survival" and endures in games that involve throwing and running (Hall, *Adolescence: Its Psychology*, 1:206). Hall's hypothesis pertained to human play, especially that of children; its application to nonhumans remains unclear.
10 Fagen, *Animal Play Behavior*, 35.
11 Brownlee, "Play in Domestic Cattle."
12 Recounted in Henig, "Taking Play Seriously."
13 Byers and Walker, "Refining the Motor Training Hypothesis."
14 Fairbanks, "Developmental Timing of Primate Play."

NOTES

15 Groos, *Play of Animals*, 81.
16 Macdonald and Sillero-Zubiri, *Biology and Conservation of Wild Canids*, 94.
17 Caro, "Effects of Experience."
18 Hall, "Object Play by Adult Animals."
19 Studies of each are cited in Barber, "Play and Energy Regulation."
20 Fagen was among those who advocated that play may have immediate benefits, helping to develop muscles and coordination.
21 Barber, "Play and Energy Regulation."
22 Thompson, "Self-Assessment in Juvenile Play."
23 Barber, "Play and Energy Regulation."
24 Ibid.
25 Thompson, "Self-Assessment in Juvenile Play."
26 Darwin, *Origin of Species*, 156.
27 Fagen and Fagen, "Juvenile Survival and Benefits" and "Play Behaviour and Multi-Year."
28 The Kalahari Meerkat Project had been established in 1993 by Tim Clutton-Brock, a behavioral ecologist at the University of Cambridge.
29 In 1982 Marc Bekoff, in "Functional Aspects of Play," had observed that designing and implementing a test of the practice hypothesis would be "a tall order."
30 Sharpe, "Play Fighting Does Not Affect."
31 Sharpe, "Play Does Not Enhance" and "Frequency of Social Play."
32 Sharpe, "Play Fighting Does Not Affect," 1023.
33 Darwin, *Origin of Species*, 14.

CHAPTER 3 Tumbling Piglets and Somersaulting Monkeys: Training for the Unexpected

1 Spinka, Newberry, and Bekoff, "Mammalian Play."
2 Ibid.
3 Biben, "Effects of Social Environment." See also Watson and Croft, "Age-Related Differences in Play-Fighting," 102, 336–46.
4 Bekoff, "Social Play Behavior."
5 Groos, *Play of Animals*, 111.
6 Pellis and Pellis, *Playful Brain*, 163–64.
7 Groos, *Play of Animals*, 114.
8 Cronin, "Muddy Baby Elephants."
9 All studies cited in Spinka, Newberry, and Bekoff, "Mammalian Play."
10 Petrů et al., "Revisiting Play Elements."
11 Ibid.

12   Loizos, "Play Behavior in Higher Primates."
13   Henig, "Taking Play Seriously."

CHAPTER 4 "Let's Go Tickle Some Rats": The Neuroscience of Play

1   "Science of the Brain," 267.
2   Pellis and Pellis, *Playful Brain*, 63.
3   Pellis, "Keeping in Touch."
4   Pellis and Pellis, *Playful Brain*, 80.
5   Einon and Morgan, "Critical Period for Social Isolation." See also Einon, Morgan, and Kibbler, "Brief Periods of Socialization."
6   Fagen used it as well. Fagan, et al., "Observing Behavioral Qualities."
7   Darwin, *Expression of the Emotions*, 186.
8   Pellis and Pellis, *Playful Brain*, 16.
9   Tinbergen, "On Aims and Methods."
10   Pellis, Pellis, and Whishaw, "Role of the Cortex."
11   Pellis et al., "Effects of Orbital Frontal Cortex Damage."
12   Bell et al., "Role of the Medial Prefrontal Cortex."
13   Brown, "Play Deprivation."
14   Ibid.
15   Gray, "Decline of Play and the Rise."
16   See, for instance, Burghardt, "Comparative Reach of Play," 353.
17   "Science of the Brain."
18   Sharpe, "So You Think You Know."
19   "Science of the Brain."
20   Sandseter and Kennair, "Children's Risky Play."
21   Pellis and Pellis, *Playful Brain*, 145.
22   "Science of the Brain."
23   See Huber, *Embracing Rough-and-Tumble Play.*
24   Burghardt, "On the origins of play," 5–41.

CHAPTER 5 Courtly Canines: Competing to Cooperate and Cooperating to Compete

1   Hare et al., "Domestication Hypothesis."
2   Hare, "Survival of the Friendliest."
3   Grimm, "How Smart Is That Doggy?"
4   Nagasawa et al., "Social Evolution."
5   Huizinga, *Homo Ludens*, 1.
6   Bekoff, "Social Communication in Canids."

7 Barber, "Play and Energy Regulation."
8 Byosiere, Espinosa, and Smuts, "Investigating the Function of Play Bows."
9 Pellis and Pellis, *Playful Brain*, 138.
10 Thierry, "Unity in Diversity."
11 Takeshita et al., "Beneficial Effect of Hot Spring Bathing."
12 Reinhart et al., "Targets and Tactics."
13 Monsó et al., "Animal Morality."

CHAPTER 6 Wood Thrush Songs, Herring Gull Drop-Catching, and Bowerbird Art;
Play as the Roots of Culture

1 Heinrich and Smolker, "Play in Common Ravens," 27–44.
2 Ibid.
3 In the late 1880s Edinger proposed that the animal brain evolved via Darwinian natural selection, but—shades of Lamarck—evolved upward on a version of Aristotle's *scala naturae*, a ranking of all living things on a scale of "perfection," proceeding from fish, to amphibians, to reptiles, to birds and mammals, to primates, and, finally—a rather anthropocentric schema to be sure—to humans. Edinger believed this evolution operated by accretion, with newer parts developing over them, as layers that provided a clear record of the brain's history. Since Edinger was the first to identify many parts of brains, it was within his purview to name them. His names reflected his estimation of their age. He believed that the larger part of the cortex—that part generally understood to make possible sophisticated, malleable, learned behavior—was the part evolved most recently. Thus he called it the *neocortex*, with the *neo* meaning "new." Edinger's model has proven to be deeply flawed. Organisms may be more complex or less complex, but in no meaningful sense is any animal brain "higher" or "lower" than any other. Moreover, most neuroscientists nowadays agree that accretion, adding new layers over old layers, is simply not how brains evolved.
4 Stacho et al., "Cortex-Like Canonical Circuit."
5 Ackerman, *Genius of Birds*, 42.
6 Olkowicz et al., "Birds Have Primate-Like Numbers."
7 In 2005 the Avian Brain Nomenclature Consortium, an international assemblage of specialists in avian, mammalian, reptilian, and fish neurobiology, developed a terminology that more accurately reflects current understanding of the avian cerebrum and its analogous structures in the brains of mammals. See Jarvis et al., "Avian Brains."

8 Watanabe et al., "Pigeons' Discrimination of Paintings."

9 Huizinga, *Homo Ludens*, 47.

10 Ficken, "Avian Play," 577.

11 Ibid.

12 Heinrich and Smolker, "Play in Common Ravens," 39.

13 Gamble and Cristol, "Drop-Catch Behaviour Is Play."

14 Diamond and Bond, *Kea, Bird of Paradox*, 77.

15 Groos, *Play of Animals*, 920.

16 Diamond and Bond, *Kea, Bird of Paradox*, 4. Conflicts with humans, habitat destruction, and poaching for the pet trade have taken their toll. Keas now comprise a small population and are "nationally endangered."

17 Ibid., 92.

18 Ibid., 80–81. In most accounts of animals engaging in play, they are doing so as participants. This report is remarkable in that the keas on the periphery seem to be engaging as spectators.

19 Pellis, "Description of Social Play."

20 https://www.natureweb.net/taxa/birds/montagusharrier.

21 Pandolfi, "Play Activity in Young Montagu's," 935–38.

22 Whiten, "Burgeoning Reach of Animal Culture."

23 Preyer, *Mental Development in the Child*, 42.

24 Groos, *Play of Animals*, 88–89.

25 Romanes, *Life and Letters of George John Romanes*, 78.

26 Darwin Correspondence Project, letter no. 12924, accessed April 8, 2022, https://www.darwinproject.ac.uk/letter/?docId=letters/DCP-LETT-12924.xml.

27 Romanes, *Animal Intelligence*, 485.

28 Huizinga, *Homo Ludens*, 173.

29 Bossley et al., "Tail Walking in a Bottlenose Dolphin."

30 Ibid.

31 Groos, *Play of Animals*, 111.

32 Warneken and Rosati, "Cognitive Capacities for Cooking."

33 Beran et al., "Chimpanzee Food Preferences." Also see Jacobs et al., "Tools and Food on Heat Lamps."

34 Haslam et al., "Primate Archaeology."

35 Ibid.

CHAPTER 7 Memes and Dreams: Dreaming as Playing without a Body

1 Godfrey-Smith, *Other Minds*, 134.

2 Pagel et al., "Definitions of Dream."

3 Flanagan, "Dreaming Is Not an Adaptation."

4 Thompson, *Passions of Animals*, 61.

5 Darwin, *Descent of Man*, 46.

6 Romanes, *Animal Intelligence*, 347.

7 Although, in the words of one researcher, "Vocalization by sleeping birds appears not yet to have received scientific attention": Malinowski, Scheel, and McCloskey, "Do Animals Dream?"

8 Ibid., 8.

9 Spinka, Newberry, and Bekoff, "Mammalian Play."

10 Ungurean et al., "Evolution and Plasticity of Sleep."

11 Although dreaming is common during REM sleep, it may occur without dreams; dreaming is less common during NREM sleep.

12 Revonsuo, "Reinterpretation of Dreams."

13 Revonsuo, Tuominen, and Vall, "Avatars in the Machine."

14 Jarvis, "Did COVID Change How We Dream?"

15 Panksepp, *Affective Neuroscience.*

16 "I woke up with a lovely tune in my head. I thought, 'That's great, I wonder what that is?' There was an upright piano next to me, to the right of the bed by the window. I got out of bed, sat at the piano, found G, found F sharp minor 7th. . . . I liked the melody a lot but because I'd dreamed it I couldn't believe I'd written it." Rybaczewski, "Beatles Music History!"

17 Wu, "Zebra Finches Dream."

18 Groos, *Play of Animals*, 310.

19 Darwin, *Descent of Man*, 60.

20 Atkinson et al., "Languages Evolve in Punctuational Bursts."

21 Darwin, *Origin of Species*, 381.

CHAPTER 8 The Evolution of Play

1 This particular evolutionary process, by which features take on functions for which they were not originally adapted, is called *exaptation*, a term coined in the 1980s by paleontologist and author Stephen J. Gould and paleontologist Elisabeth S. Vrba. The portmanteau word is from *ex*, Latin for "out of," and *aptation*, a modification of *adaptation*.

2 All from Pellis and Pellis, *Playful Brain.*

3 This is true only with healthy rats; recall the Pellises and McKenna's discovery that a rat whose stratium had been compromised will confuse offensive and defensive moves.

4 Pellis and Pellis, "What Is Play Fighting?," 355–66.

5 Guéguen, "Men's Sense of Humor," 145–56.

6 Croft and Snaith, "Boxing in Red Kangaroos." A third mammalian order, called monotremes, has two extant families, the duck-billed platypus and the spiny anteater. No one has unequivocally identified play in either, although there are anecdotal accounts of captive platypuses playing like puppies. Bennett, "Notes on the Natural History."

7 Nowak, 1999.

8 Lamarck's metaphorical ladder of phyla through ever-more-complex forms is a stubbornly persistent idea and survives in our vocabulary. In some circles, mammals are still sometimes termed "higher" animals.

9 Burghardt, "Comparative Reach of Play."

10 Burghardt, *Genesis of Animal Play*, 297.

11 Burghardt, Ward, and Rosscoe, "Problem of Reptile Play."

12 Burghardt, *Genesis of Animal Play*, 284–89.

13 And, alas, another species that the International Union for Conservation of Nature identifies as "threatened."

14 Incidentally, a reason bottlenose dolphins and seals are among the most playful mammals is that they enjoy both high metabolisms and aquatic environments.

15 Jarmer, *Das Seelenleben der Fische.*

16 All cited in Burghardt, *Genesis of Animal Play*, 313–57.

17 Ibid., 339.

18 Ibid., 321.

19 Ibid., 329.

20 Ibid., 340–42.

21 Hölldobler and Wilson, *Ants*, 370. Darwin noted, "Even insects play together, as has been described by that excellent observer, P. Huber, who saw ants chasing and pretending to bite each other, like so many puppies" (*Descent of Man*, 39). Pierre Huber was a much-respected naturalist, yet many think in this particular observation he may have been in error. Hölldobler and Wilson suggested that what Huber had witnessed was probably a struggle between two colonies.

22 Galpayage Dona et al., "Do Bumble Bees Play?"

23 Dapporto et al., "Dominance Interactions in Young Adult."

24 Hildenbrand et al., "Potential Cephalopod."

25 Graham and Burghardt, "Current Perspectives on the Biological Study."

26 Tzar and Scigliano, "Through the Eye of an Octopus."

27 Yoshida, Yura, and Ogura, "Cephalopod Eye Evolution."

CHAPTER 9 Innovative Gorillas: The Surprising Role of Play in Natural Selection

1 Pellis and Pellis, *Playful Brain*, 126.

2 Darwin, *Origin of Species*, 420.

3 Darwin, *Expression of the Emotions*, xviii. Darwin's *On the Origin of Species*, published in 1859, hinted that "light would be thrown on "the origin of man and his history," but did not explicitly state that humans were a product of the same forces that produced, say, orangutans.

4 Darwin, *Origin of Species*, 60.

5 In the first decades of the twentieth century, Mendelian genetics, explaining the mechanism of natural selection, was fused with Darwin's theory to yield the unified theory of evolution called the Modern Synthesis. In the Modern Synthesis, an organism's evolutionary success is measured not in number of individual descendants, but the number of genes passed on to those descendants. By the late twentieth century, the Modern Synthesis, too, was proving to have shortcomings. Attempts to address them came, and continue to come, from fields of developmental biology, epigenetics, genomics, horizontal gene transfer, molecular biology, microbiology, and symbiogenesis.

6 Groos, *Play of Animals*, xxi. Darwin and his advocates were able to answer each critique of natural selection or to dismiss them as unimportant. The objection that explanations for advantageous variations could be ad hoc were made, and are still being made, by those who haven't read *Origin*. In fact, Darwin rarely employs ad hoc explanations for characteristics of individual organisms. As many noted, he does offer explanations in the manner of a rigorous theory—for instance correlating patterns in one species' morphology with its geographical distribution to make a testable prediction about the same patterns in another species. The objection that Darwin's estimate of the span of time necessary for the evolution of earth's life was greater than the lifetime of the sun was based upon an imperfect nineteenth-century understanding of physics, but by that measure it was valid. In a note to his friend Joseph Hooker, Darwin acknowledged it as such, writing, "Some of the remarks about the lapse of years are very good, & the Reviewer gives me some good & well deserved raps,—confound it I am sorry to confess the truth." In the sentence following Darwin pronounced the objection irrelevant. "But," he continued, "it does not at all concern main argument." In the third and later editions of *Origin* he honed that argument considerably and simply removed all numerical references to geological time spans. (He would be careful

to include none in the 1871 publication of *The Descent of Man*.) As for the objection that natural selection did not explain variation, Darwin had been fully aware of the shortcoming. He hoped that he would eventually discover an explanation for variability and heritability; in the meantime, he believed, it was enough to show that they occurred.

7 In 1972 paleontologists Niles Eldridge and Stephen Jay Gould suggested that those gaps may be explained as products of *punctuated equilibria*, a phenomenon by which long stretches of gradual development are interrupted by brief episodes of rapid change.

8 Darwin, *Origin of Species*, 15.

9 Ibid., 119.

10 Weismann, *Das Keimplasma: Eine Theorie der Vererbung*.

11 Darwin, *Origin of Species*, 362.

12 Darwin's religious sentiments changed during his lifetime. In an 1879 letter, when he was seventy, he wrote, "I have never been an atheist in the sense of denying the existence of God. – I think that generally (& more and more so as I grow older) but not always, that an agnostic would be the most correct description of my state of mind." Ibid., 384n.

13 Ibid., 414.

14 "I venture to suggest that the theory which regards play as a native tendency of an animal to practice certain functions, before they are required of him, be called the 'practice theory' of play." Groos, *Play of Animals*, vii.

15 Wozniak and Santiago-Blay, "Trouble at Tyson Alley."

16 Baldwin, "New Factor in Evolution."

17 It's mildly curious that Baldwin in his appendix to *The Play of Animals* did not engage that text. That is, he did not connect or try to connect Groos's ideas of animal play with organic selection. In 1901 he provided the preface to Groos's companion work for *The Play of Animals*, *The Play of Man*, and a year later Baldwin contributed the entry on play in *Dictionary of Philosophy and Psychology*. In neither piece did he reference organic selection. For his part, Groos likewise failed to make a connection. In *The Play of Animals*, he made no mention of Baldwin's idea; although he alluded to it in *The Play of Man*, he did not relate it to play. It was as though two strangers, seated side by side on a long train ride, found that they had much in common, but when the train arrived at its destination, they amicably parted ways, neither realizing that they had failed to hit upon the subject that might have been most interesting to each.

18 In his 1909 work *Darwin and the Humanities*, Baldwin argues (p. 19) that Darwin himself allowed for organic selection.

19 Wozniak and Santiago-Blay, "Trouble at Tyson Alley."

20 Huxley, *Evolution.*

21 Simpson, "Baldwin Effect."

22 Yeh and Price, "Adaptive Phenotypic Plasticity."

23 Badyaev, Alexander V. "Evolutionary Significance of Phenotypic Accommodation."

24 Podlipniak, "Role of the Baldwin Effect."

25 Wilson and Kleiman, "Eliciting Play." Also Lund and Vestergaard, "Development of Social Behaviour."

CHAPTER 10 Playing Animal

1 Morell, *Animal Wise*, 49.

2 Burghardt, "Amending Tinbergen."

3 Since in Christian doctrine the soul is immortal, an animal endowed with a soul might have a place in the afterlife. The Bible's positions on the question, not surprisingly, are contradictory and many. Pressed for a favorite, I'd take Ecclesiastes's characteristically humble admission: "Who knoweth the spirit of man that goeth upward, and the spirit of the beast that goeth downward to the earth?"

4 Salisbury, "Do Animals Go to Heaven?"

5 Hume, *Treatise of Human Nature*, 176.

6 Thompson, *Passions of Animals*, 61.

7 Darwin, *Descent of Man*, 48–49.

8 Romanes, *Animal Intelligence*, 10.

9 Ibid., viii.

10 Morgan, *Introduction to Comparative Psychology*, 53. This is the same Morgan who in two years would concurrently with (but independently of) Henry Fairfield Osborn and James Mark Baldwin propose the hypothesis that Baldwin termed organic selection.

11 Groos, *Play of Animals*, 83.

12 *Oxford Companion to Animal Behavior*, 55.

13 de Waal, "Silent Invasion."

14 Ackerman, *Genius of Birds*, 34.

15 Eiseley, *Star Thrower*, 30.

16 Singer, "Bat Echoes Used as Virtual Reality Guide."

17 Yuqian Ma et al., "Mammalian Near-Infrared Image Vision."

18 Washburn, *Animal Mind*, 44.

19 Not all who inhabit the consciousness and body of an animal—or believe they do—control that experience. In species dysphoria one believes that one's body is from the wrong species.

20 Brulliard, "This Man Lived as a Goat."

21 Thwaites, *GoatMan*, 176.

22 Foster, *Being a Beast*, 54.

23 Ibid., 48.

24 Pilcher, "Man Who Lived like a Goat."

25 Ibid.

26 Foster, *Being a Beast*, 124.

27 Ibid., 193.

28 Maglieri et al., "Levelling Playing Field."

EPILOGUE Play, Life, and Everything

1 Dewey, *Influence of Darwin*, 16.

Bibliography

Aberth, John. *An Environmental History of the Middle Ages: The Crucible of Nature*. New York: Routledge, 2013.

Ackerman, Jennifer. *The Genius of Birds*. New York: Penguin Books, 2017.

Atkinson, Quentin D., et al. "Languages Evolve in Punctuational Bursts." *Science* 319, no. 5863 (2008): 588. DOI:10.1126/science.1149683.

Ayala, Francisco José, and Andrey Rzhetsky. "Origin of the Metazoan Phyla: Molecular Clocks Confirm Paleontological Estimates." *Proceedings of the National Academy of Sciences* 95, no. 2 (January 1998): 606–11. DOI:10.1073/pnas.95.2.606.

Badyaev, Alexander V. "Evolutionary Significance of Phenotypic Accommodation in Novel Environments: An Empirical Test of the Baldwin Effect." *Philosophical Transactions of the Royal Society of London. Series B, Biological Sciences* 364, no. 1520 (2009): 1125–41. DOI:10.1098/rstb.2008.0285.

Baldwin, James Mark. *Darwin and the Humanities*. London: American Mathematical Society, 1909.

———. "A New Factor in Evolution." *The American Naturalist* 30, no. 354 (1896): 441–51.

Barber, N. "Play and Energy Regulation in Mammals." *Quarterly Review of Biology* 66 (1991): 129–47.

Bekoff, M. "Functional Aspects of Play as Revealed by Structural Components and Social Interaction Patterns." *Behavioral and Brain Sciences* 5 (1982): 156–57.

———. "Social Communication in Canids, Evidence for the Evolution of a Stereotyped Mammalian Display." *Science* 197 (1977): 1097–99.

———. "Social Play Behavior." *Bioscience* 34, no. 4 (1984): 228–33.

Bekoff, M., and C. Allen. "Intentional Communication and Social Play: How and Why Animals Negotiate and Agree to Play." In *Animal Play:*

Evolutionary, Comparative and Ecological Perspectives, edited by Mark Bekoff and John A. Byers, 161–82. Cambridge: Cambridge University Press, 1998.

Bekoff, Marc, and John A. Byers, eds. *Animal Play: Evolutionary, Comparative and Ecological Perspectives*. Cambridge: Cambridge University Press, 1998.

———. "A Critical Reanalysis of the Ontogeny and Phylogeny of Mammalian Social and Locomotor Play: An Ethological Hornet's Nest." In *Behavioral Development: The Bielefeld Interdisciplinary Project*, edited by K. Immelmann, G. W. Barlow, L. Petrinovich, and M. Main, 296–337. Cambridge: Cambridge University Press, 1981.

Bell, Heather C., et al. "The Role of the Medial Prefrontal Cortex in the Play Fighting of Rats." *Behavioral Neuroscience* 123 (2009): 1158–68.

Bennett, George. "Notes on the Natural History and Habits of the *Ornithorhynchus paradoxus*, Blum." *Transactions of the Zoological Society of London* 1, no. 3 (1835): 229–58.

Beran, Michael J., et al. "Chimpanzee Food Preferences, Associative Learning, and the Origins of Cooking." *Learning & Behavior* 44, no. 2 (2016): 103–8.

Biben, M. "Effects of Social Environment on Play in Squirrel Monkeys (*Saimiri sciureus*): Resolving Harlequin's Dilemma." *Ethology* 81 (1989): 72–82.

Boal, Jean Geary, Andrew W. Dunham, Kevin T. Williams, and Roger T. Hanlon. "Experimental Evidence for Spatial Learning in Octopuses (*Octopus bimaculoides*)." *Journal of Comparative Psychology* 114, no. 3 (2000): 251.

Borrell, Brendan. "Are Octopuses Smart? The Mischievous Mollusk That Flooded a Santa Monica Aquarium Is Not the First MENSA-Worthy Octopus." Interview with Jennifer Mather. *Scientific American*, February 27, 2009.

Bossley, Mike, et al. "Tail Walking in a Bottlenose Dolphin Community: The Rise and Fall of an Arbitrary Cultural 'Fad.'" *Biology Letters* 14 (2018).

Brown, Stuart. "Play Deprivation . . . a Leading Indicator for Mass Murder." June 1, 2014. http://www.nifplay.org/play-deprivation-a-leading-indicator-for-mass-murder/.

Brownlee, A. "Play in Domestic Cattle in Britain: An Analysis of Its Nature." *British Veterinary Journal* 110 (1954).

Brulliard, Karin. "This Man Lived as a Goat for Nearly a Week. We Asked Him Why." Https://www.washingtonpost.com. May 25, 2016.

Burghardt, Gordon M. "Amending Tinbergen: A Fifth Aim for Ethology." In *Anthropomorphism, Anecdotes and Animals*, edited by Robert W. Mitchell et al. Albany: State University of New York, 1997.

———. "The Comparative Reach of Play and Brain: Perspective, Evidence, and Implications." *American Journal of Play* 2 (2010): 338–56.

———. *The Genesis of Animal Play: Testing the Limits*. Cambridge, MA: Bradford Books, MIT Press, 2005.

———. "On the Origins of Play." In *Play in Animals and Humans*, edited by Peter K. Smith, 5–41. Hoboken, NJ: Blackwell, 1984.

Burghardt, Gordon M., Brian Ward, and Roger Rosscoe. "Problem of Reptile Play: Environmental Enrichment and Play Behavior in a Captive Nile Soft-Shelled Turtle, *Trionyx triunguis*." *Zoo Biology* 15 (1996): 223–38.

Byers, J. A. "Terrain Preferences in the Play Behaviour of Siberian Ibex Kids." *Zeitschrift für Tierpsychologie* 45 (1977): 199–209.

Byers, J. A., and C. Walker. "Refining the Motor Training Hypothesis for the Evolution of Play." *American Naturalist* 146 (1995): 25–40.

Byosiere, S. E., J. Espinosa, and B. Smuts. "Investigating the Function of Play Bows in Adult Pet Dogs (*Canis lupus familiaris*)." *Behavioural Processes* 125 (2016): 106–13. PMID:26923096.

Caro, T. M. "The Effects of Experience on the Predatory Patterns of Cats." *Behavioral and Neural Biology* 29 (1980): 1–28.

Carroll, Sean B., Jennifer K. Grenier, and Scott D. Weatherbee. *From DNA to Diversity: Molecular Genetics and the Evolution of Animal Design*. 2nd ed. Hoboken, NJ: Wiley-Blackwell, 2004.

Chi, Z., and D. Margoliash. "Temporal Precision and Temporal Drift in Brain and Behavior of Zebra Finch Song." *Neuron* 32, no. 5 (2001): 899–910. DOI:10.1016/s0896-6273(01)00524-4. PMID:11738034.

Courage, Katherine Harmon. *Octopus!: The Most Mysterious Creature in the Sea*. New York: Current, 2013.

Croft, D. B., and F. Snaith. "Boxing in Red Kangaroos, *Macropus rufus*: Aggression or Play?" *International Journal of Comparative Psychology* 4, no. 3 (1990). DOI:10.5070/P443013956.

Cronin, Melissa. "Muddy Baby Elephants Play Slip 'n Slide Like Pros." *Daily Dodo*, January 12, 2015.

Cross, Craig. *The Beatles: Day-by-Day, Song-by-Song, Record-by-Record*. Lincoln, NE: iUniverse, 2005.

Dapporto, Leonardo, et al. "Dominance Interactions in Young Adult Paper Wasp (*Polistes dominulus*) Foundresses: A Playlike Behavior?" *Journal of Comparative Psychology* 120, no. 4 (2006): 394–400.

Darwin, Charles. *The Descent of Man, and Selection in Relation to Sex*. 2nd rev. and augmented ed. London: John Murray, 1877.

———. *The Expression of the Emotions in Man and Animals*. Edited by Joe Cain and Sharon Messenger, with an introduction by Joe Cain. London: Penguin Classics, 2009. From the 2nd ed., 1890.

———. *The Origin of Species*. Introduction and notes by George Levine. New York: Barnes & Noble Books, 2004.

Deutsch, David. *The Beginning of Infinity: Explanations That Transform the World*. New York: Penguin Books, 2012.

de Waal, F. "Silent Invasion: Imanishi's Primatology and Cultural Bias in Science." *Animal Cognition* 6 (2003): 293–99.

Dewey, John. *The Influence of Darwin on Philosophy, and Other Essays in Contemporary Thought*. New York: Henry Holt, 1910.

Diamond, Judy, and Alan B. Bond. *Kea, Bird of Paradox: The Evolution and Behavior of a New Zealand Parrot*. Berkeley and Los Angeles: University of California Press, 1999.

Dixon, Roland Burrage. *Oceanic Mythology*. Boston: Marshall Jones, 1916.

Dobzhansky, Theodosius. "Nothing in Biology Makes Sense Except in the Light of Evolution." *American Biology Teacher* 35, no. 3 (March 1973): 125–29.

Edinger, L. *Investigations on the Comparative Anatomy of the Brain*. 5 vols. Frankfurt/Main: Moritz Diesterweg, 1888–1903. Translation from German.

Einon, Dorothy F., and Michael J. Morgan. "A Critical Period for Social Isolation in the Rat." *Developmental Psychobiology* 10 (1977): 123–32.

Einon, Dorothy F., Michael J. Morgan, and Christopher C. Kibbler. "Brief Periods of Socialization and Later Behavior in the Rat." *Developmental Psychobiology* 11 (1978): 213–25.

Eiseley, Loren C. *The Star Thrower*. New York: Harvest Books, 1979.

Fagan, R., J. Conitz, and E. Kunibe. "Observing Behavioral Qualities." *International Journal of Comparative Psychology* 10, no. 4 (1997). Http:// dx.doi.org/10.46867/C4BP41. Retrieved from https://escholarship .org/uc/item/8gx671cb.

Fagen, R. M. *Animal Play Behavior*. New York: Oxford University Press, 1981.

Fagen, Robert, and Johanna Fagen. "Juvenile Survival and Benefits of Play Behaviour in Brown Bears, *Ursus arctos*." *Evolutionary Ecology Research* 6 (2004): 89–102.

———. "Play Behaviour and Multi-Year Juvenile Survival in Free-Ranging Brown Bears, *Ursus arctos*." *Evolutionary Ecology Research* 11 (2009): 1–15.

Fairbanks, L. A. "The Developmental Timing of Primate Play: A Neural Selection Model." In *Biology, Brains, and Behavior: The Evolution of Human Development*, edited by S. T. Parker, J. Langer, and M. L. McKinney, 131–58. Santa Fe, NM: School of American Research Press, 2000.

BIBLIOGRAPHY

Ficken, Millicent S. "Avian Play." *Auk* 94, no. 3 (July 1, 1977): 573–82. DOI:10.1093/auk/94.3.573.

Flanagan, O. "Dreaming Is Not an Adaptation." In *Sleep and Dreaming: Scientific Advances and Reconsiderations*, edited by E. Pace-Schott, M. Solms, M. Blagrove, and S. Harnad, 936–39. New York: Cambridge University Press, 2003.

Foster, Charles. *Being a Beast: Adventures across the Species Divide.* New York: Holt, 2016.

Galpayage Dona, Hiruni Samadi, et al., "Do Bumble Bees Play?" *Animal Behaviour* 194 (2022): 239–51.

Gamble, Jennifer R., and Daniel A. Cristol. "Drop-Catch Behaviour Is Play in Herring Gulls, *Larus argentatus*." *Animal Behaviour* 63, no. 2 (2002). DOI:10.1006/anbe.2001.1903.

Gill, Frank B. *Ornithology.* 3rd ed. National Audubon Society. New York: W. H. Freeman, 2006.

Gladstone, Rick. "Dogs in Heaven? Pope Francis Leaves Pearly Gates Open." *New York Times*, December 11, 2014.

Godfrey-Smith, Peter. *Other Minds: The Octopus, the Sea, and the Deep Origins of Consciousness.* New York: Farrar, Straus and Giroux, 2016.

Goode, Erica. "Learning from Animal Friendships." *New York Times*, January 26, 2015.

Gould, Stephen Jay, and Elisabeth S. Vrba. "Exaptation—a Missing Term in the Science of Form." *Paleobiology* 8, no.1 (Winter 1982): 4–15.

Graham, K. L., and G. M. Burghardt. "Current Perspectives on the Biological Study of Play: Signs of Progress." *Quarterly Review of Biology* 85, no. 4 (December 2010): 393–418. DOI:10.1086/656903. PMID:21243962.

Gray, Peter. "The Decline of Play and the Rise of Psychopathology in Children and Adolescents." *American Journal of Play* 3 (2011): 443–63.

Griffin, D. R. *Animal Thinking.* Cambridge, MA: Harvard University Press, 1984.

Grimm, David. "How Smart Is That Doggy in the Window?" *Time*, April 13, 2014.

Groos, Karl. *The Play of Animals.* Translated by Elizabeth Baldwin. Preface and an appendix by J. Mark Baldwin. New York: D. Appleton, 1898.

Guéguen, Nicolas. "Men's Sense of Humor and Women's Responses to Courtship Solicitations: An Experimental Field Study." *Psychological Reports* 107 (2010): 145–56.

Hall, G. S. *Adolescence: Its Psychology and Its Relations to Physiology, Anthropology, Sociology, Sex, Crime, Religion and Education.* Vol. 1. New York: D. Appleton, 1904. DOI:10.1037/10616-000.

Hall, Sarah L. "Object Play by Adult Animals." In *Animal Play: Evolutionary, Comparative and Ecological Perspectives,* edited by Mark Bekoff and John A. Byers, 45–60. Cambridge: Cambridge University Press, 1998.

Harcourt, R. "Survivorship Costs of Play in the South American Fur Seal." *Animal Behaviour* 42, no. 3 (1991): 509–11. DOI:10.1016/S0003-3472(05)80055-7.

Hare, Brian. "Survival of the Friendliest: *Homo sapiens* Evolved via Selection for Prosociality." *Annual Review of Psychology* 68 (2017): 155–86.

Hare B., et al. "The Domestication Hypothesis for Dogs' Skills with Human Communication: A Response to Udell et al. 2008 and Wynne et al. 2008." *Animal Behaviour* 79 (2010): e1–e6.

Haslam, M., et al. "Primate Archaeology." *Nature* 460 (2009): 339–44. DOI:10.1038/nature08188.

Hausfater, G. "Predatory Behavior of Yellow Baboons." *Behaviour* 56 (1976): 44–68.

Heinrich, Bernd. "An Experimental Investigation of Insight in Common Ravens (*Corvus corax*)." *Auk* 112, no. 4 (1995): 994–1003. www.jstor.org/stable/4089030.

Heinrich, Bernd, and Rachel Smolker. "Play in Common Ravens (*Corvus corax*)." In *Animal Play: Evolutionary, Comparative and Ecological Perspectives,* edited by Mark Bekoff and John A. Byers, 27–44. Cambridge: Cambridge University Press, 1998.

Henig, Robin Marantz. "Taking Play Seriously." *New York Times Magazine,* February 17, 2008.

Hildenbrand, Anne, et al. "A Potential Cephalopod from the Early Cambrian of Eastern Newfoundland, Canada." *Communications Biology* 4, no. 1 (2021). DOI:10.1038/s42003-021-01885-w.

Himmler, Stephanie M., et al. "Play, Variation in Play and the Development of Socially Competent Rats." *Behaviour* 153 (2016): 1103–37.

Hölldobler, Bert, and Edward O. Wilson. *The Ants.* Cambridge, MA: Belknap Press of Harvard University Press, 1990.

Horváth, Zsuzsánna, Antal Dóka, and Adám Miklósi. "Affiliative and Disciplinary Behavior of Human Handlers during Play with Their Dog Affects Cortisol Concentrations in Opposite Directions." *Hormones and Behavior* 54, no. 1 (June 2008): 107–14. DOI:10.1016/j.yhbeh.2008.02.002.

Huber, Mike. *Embracing Rough-and-Tumble Play: Teaching with the Body in Mind.* St. Paul, MN: Redleaf Press, 2016.

Huizinga, J. *Homo Ludens: A Study of the Play-Element in Culture.* London: Routledge & Kegan Paul, 1949.

Hume, David. *A Treatise of Human Nature.* Oxford: Clarendon Press, 1739.

BIBLIOGRAPHY

Huxley, Julian. *Evolution: The Modern Synthesis.* London: George Allen & Unwin, 1942.

Hyland, D. A. *The Question of Play.* Lanham, MD: University Press of America, 1984.

Ikemoto, Satoshi, and Jaak Panksepp. "The Effects of Early Social Isolation on the Motivation for Social Play in Juvenile Rats. *Developmental Psychobiology* 25 (1992): 261–74.

Immelmann, Klaus, and Colin Beer. *A Dictionary of Ethology.* Cambridge, MA: Harvard University Press, 1989.

Irwin, Aisling. "First Warm-Blooded Lizards Switch on Mystery Heat Source at Will." *New Scientist,* January 22, 2016.

Jacobs, Ivo F., et al. "Tools and Food on Heat Lamps: Pyrocognitive Sparks in New Caledonian Crows?" *Behaviour* 159, no. 6 (2021).

Jarmer, Karl. *Das Seelenleben der Fische.* Munich: R. Oldenbourg, 1928.

Jarvis, Brooke. "Did COVID Change How We Dream?" *New York Times Magazine,* November 3, 2021.

Jarvis, E., et al. "Avian Brains and a New Understanding of Vertebrate Brain Evolution." *Nature Reviews Neuroscience* 6 (2005): 151–59. DOI:10.1038/nrn1606.

Knoper, Randall. *Literary Neurophysiology: Memory, Race, Sex, and Representation in U.S. Writing, 1860–1914.* New York: Oxford University Press, 2021.

Kuba, M., et al. "Looking at Play in *Octopus vulgaris.*" *Berliner Paläontologische Abhandlungen* 3 (2003): 163–69.

Kuba, M. J., R. A. Byrne, D. V. Meisel, and J. A. Mather. "When Do Octopuses Play? Effects of Repeated Testing, Object Type, Age, and Food Deprivation on Object Play in *Octopus vulgaris.*" *Journal of Comparative Psychology* 120 (2006): 184–90.

Laland, Kevin N., and Vincent M. Janik. "The Animal Cultures Debate." *Trends in Ecology & Evolution* 21, no. 10 (2006): 542–47.

Leslie, A. M. "Pretense and Representation. The Origins of 'Theory of Mind.'" *Psychological Review* 94 (1987): 412–26.

Lewes, G. H. *Seaside Studies at Ilfracombe, Tenby, the Scilly Isles, and Jersey.* 2nd ed. Edinburgh: William Blackwood & Sons, 1860.

Lincoln, Jackson S. *The Dream in Native American and Other Primitive Cultures.* Mineola, NY: Dover Publications, 2003.

Loizos, C. "Play Behavior in Higher Primates: A Review." In *Primate Ethology,* edited by D. Morris, 176–218. Chicago: Aldine, 1967.

Lund, D., and K. S. Vestergaard. "Development of Social Behaviour in Four Litters of Dogs (*Canis familiaris*)." *Acta Veterinaria Scandinavica* 39 (1998): 183–93.

Macdonald, David W., and Claudio Sillero-Zubiri. *The Biology and Conservation of Wild Canids.* New York: Oxford University Press, 2004.

Maglieri, Veronica, Filippo Bigozzi, Marco Germain Riccobono, and Elisabetta Palagi. "Levelling Playing Field: Synchronization and Rapid Facial Mimicry in Dog-Horse Play." *Behavioural Processes* 174 (2020).

Malinowski, J. E., D. Scheel, and M. McCloskey. "Do Animals Dream?" *Consciousness and Cognition* 95 (2021). DOI:10.1016/j.concog.2021.103214.

Marek, Roger, Cornelia Strobel, Timothy W. Bredy, and Pankaj Sah. "The Amygdala and Medial Prefrontal Cortex: Partners in the Fear Circuit." *Journal of Physiology* 591, pt. 10 (2013): 2381–91. DOI:10.1113/jphysiol.2012.248575.

Martin, P., and T. Caro. "On the Functions of Play and Its Role in Behavioral Development." *Advances in the Study of Behavior* 15 (1985): 59–103.

Marzluff, J. M., B. Heinrich, and C. S. Marzluff. "Raven Roosts Are Mobile Information Centres." *Animal Behavior* 51 (1996): 89–103.

Mason, G. J. "Stereotypies: A Critical Review." *Animal Behaviour* 41 (1991): 101–37. DOI:10.1016/S0003-3472(05)80640-2.

Mather, Jennifer. "Octopuses Are Smart Suckers!?" Cephalopod Page. http://www.thecephalopodpage.org/smarts.php.

Mather, Jennifer A., and Roland C. Anderson. "Exploration, Play and Habituation in Octopuses (*Octopus dofleini*)." *Journal of Comparative Psychology* 113 (1999): 333–38.

Miklosi, A., et al. "A Simple Reason for a Big Difference: Wolves Do Not Look Back at Humans, but Dogs Do." *Current Biology* 13 (2003): 763–66.

Monsó, Susana, et al. "Animal Morality: What It Means and Why It Matters." *Journal of Ethics* 22, no. 3 (2018).

Montgomery, Sy. *The Soul of an Octopus: A Surprising Exploration into the Wonder of Consciousness.* New York: Atria Books, 2015.

Morell, Virginia. *Animal Wise: How We Know Animals Think and Feel.* New York: Broadway Books, 2013.

Morgan, C. Lloyd. *An Introduction to Comparative Psychology.* London: Walter Scott, 1894.

Nagasawa, Miho, et al. "Social Evolution. Oxytocin-Gaze Positive Loop and the Coevolution of Human-Dog Bonds." *Science* 348, no. 6232 (2015): 333–36. DOI:10.1126/science.1261022.

Newberry, R. C., D. G. M. Wood-Gush, and J. W. Hall. "Playful Behaviour of Piglets." *Behavioural Processes* 17 (1988): 205–16.

Olkowicz, Seweryn, et al. "Birds Have Primate-Like Numbers of Neurons in the Forebrain." *Proceedings of the National Academy of Sciences,* June 13, 2016. DOI:10.1073/pnas.1517131113.

Pagel, J., et al. "Definitions of Dream: A Paradigm for Comparing Field Descriptive Specific Studies of Dream." *Dreaming* 11 (2001): 195–202. DOI:10.1023/A:1012240307661.

Pandolfi, Massimo. "Play Activity in Young Montagu's Harriers (*Circus pygargus*)." *Auk* 113, no. 4 (1996): 935–38. Www.jstor.org/stable/4088874.

Panksepp, Jaak. *Affective Neuroscience.* New York: Oxford University Press, 1998.

Paquette, Daniel. "Fighting and Playfighting in Captive Adolescent Chimpanzees." *Aggressive Behavior* 20 (1994): 49–65.

Patton, Paul. "Ludwig Edinger: The Vertebrate Series and Comparative Neuroanatomy." *Journal of the History of the Neurosciences* 24, no. 1 (2015): 26–57. DOI:10.1080/0964704X.2014.917251.

Pellis, Sergio M. "A Description of Social Play by the Australian Magpie *Gymnorhina tibicen* Based on Eshkol-Wachman Notation." *Bird Behavior* 3, no. 3 (1981): 61–79. DOI:10.3727/015613881791560685.

———. "Keeping in Touch: Play Fighting and Social Knowledge." In *The Cognitive Animal: Empirical and Theoretical Perspectives on Animal Cognition,* edited by M. Bekoff, C. Allen, and G. M. Burghardt, 421–27. Cambridge, MA: MIT Press, 2002.

Pellis, Sergio M., et al. "The Effects of Orbital Frontal Cortex Damage on the Modulation of Defensive Responses by Rats in Playful and Nonplayful Social Contexts. *Behavioral Neuroscience* 120 (2006): 72–84.

Pellis, Sergio, and Vivien Pellis. *The Playful Brain: Venturing to the Limits of Neuroscience.* 1st ed. London: Oneworld Publications, 2010.

———. "What Is Play Fighting and What Is It Good For?" *Learning & Behavior* 45 (2017): 355–66.

Pellis, Sergio M., Vivien C. Pellis, and Mario M. McKenna. "Some Subordinates Are More Equal Than Others: Play Fighting amongst Adult Subordinate Male Rats." *Aggressive Behavior* 19 (1993): 385–93.

Pellis, S. M., V. C. Pellis, and C. J. Reinhart. "The Evolution of Social Play." In *Formative Experiences: The Interaction of Caregiving, Culture, and Developmental Psychobiology,* edited by C. M. Worthman, P. M. Plotsky, D. S. Schechter, and C. A. Cummings, 404–31. Cambridge: Cambridge University Press, 2010. DOI:10.1017/CBO9780511711879.037.

Pellis, Sergio M., Vivien C. Pellis, and Ian Q. Whishaw. "The Role of the Cortex in Play Fighting by Rats: Developmental and Evolutionary Implications." *Brain, Behavior and Evolution* 39 (1992): 270–84.

Petrů, M., M. Spinka, V. Charvátová, and S. Lhota. "Revisiting Play Elements and Self-Handicapping in Play: A Comparative Ethogram of Five Old World Monkey Species." *Journal of Comparative Psychology* 123, no. 3 (2009): 250–63. DOI:10.1037/a0016217. PMID:19685966.

Pilcher, Helen. "The Man Who Lived like a Goat." *BBC Science Focus Magazine*, February14, 2017.

Piqueret, Baptiste, et al. "Ants Learn Fast and Do Not Forget: Associative Olfactory Learning, Memory and Extinction in *Formica fusca*." *Royal Society Open Science* 6 (2019).

Podlipniak, Piotr. "The Role of the Baldwin Effect in the Evolution of Human Musicality." *Frontiers in Neuroscience* 11, no. 542 (2017). DOI:10.3389/fnins.2017.00542.

Ponomarev, L. I., and I. V. Kurchatov. *The Quantum Dice*. 2nd ed. Moscow: CRC Press, 1993.

Preyer, W. *Mental Development in the Child*. Translated by H. W. Brown. New York: Appleton, 1893.

Reinhart, C. J., et al. "Targets and Tactics of Play Fighting: Competitive *versus* Cooperative Styles of Play in Japanese and Tonkean Macaques." *International Journal of Comparative Psychology* 23 (2010): 166–200.

Revonsuo, Antti. "The Reinterpretation of Dreams: An Evolutionary Hypothesis of the Function of Dreaming." *Behavioral and Brain Sciences* 23, no. 6 (2000): 877–901, 904–1018, 1083–121. DOI:10.1017/S0140525X00004015.

Revonsuo, A., J. Tuominen, and K. Vall. "Avatars in the Machine: Dreaming as a Simulation of Social Reality." In *Open MIND: Philosophy of Mind and the Cognitive Sciences in the 21st Century*, 1st ed., vol. 2, edited by T. Metzinger and J. Windt, 1295–322. Cambridge, MA: MIT Press, 2016.

Romanes, George John. *Animal Intelligence*. 4th ed. London: Kegan Paul, Trench, 1886.

———. *The Life and Letters of George John Romanes*. Edited by Ethel Duncan Romanes. Cambridge: Cambridge University Press, 2011.

———. *Mental Evolution in Animals*. London: Kegan Paul, Trench, Trübner, 1893.

Rybaczewski, Dave. "Beatles Music History! The In-Depth Story behind the Songs of the Beatles!" http://www.beatlesebooks.com.

Salisbury, Joyce E. "Do Animals Go to Heaven? Medieval Philosophers Contemplate Heavenly Human Exceptionalism." *Athens Journal of Humanities & Arts* 1, no. 1 (2013): 79–86.

Sanders, N., and D. Gordon. "Resources and the Flexible Allocation of Work in the Desert Ant, *Aphaenogaster cockerelli*." *Insectes sociaux* 49 (2002): 371–79. DOI:org.silk.library.umass.edu/10.1007/PL00012661.

Sandseter, Ellen Beate Hansen, and Leif Edward Ottesen Kennair. "Children's Risky Play from an Evolutionary Perspective: The Anti-Phobic Effects of Thrilling Experiences." *Evolutionary Psychology* 9 (2011).

Schiller, Friedrich. *On the Aesthetic Education of Man.* Translated by E. M. Wilkinson and L. A. Willoughby. Oxford: Oxford University Press, 1967.

Schipani, Sam. "The History of the Lab Rat Is Full of Scientific Triumphs and Ethical Quandaries." Smithsonian.com, February 27, 2019.

"Science of the Brain as a Gateway to Understanding Play: An Interview with Jaak Panksepp." *American Journal of Play,* Winter 2010.

Sharpe, L. L. "Frequency of Social Play Does Not Affect Dispersal Partnerships in Wild Meerkats." *Animal Behaviour* 70 (2005): 559–69.

———. "Play Does Not Enhance Social Cohesion in a Cooperative Mammal." *Animal Behaviour* 70 (2005): 551–58.

———. "Play Fighting Does Not Affect Subsequent Fighting Success in Wild Meerkats." *Animal Behaviour* 69 (2005): 1023–29.

———. "Play and Social Relationships in the Meerkat (*Suricata suricatta*)." PhD diss., Stellenbosch University, 2005.

———. "So You Think You Know Why Animals Play . . ." Guest blog. *Scientific American,* May 17, 2011.

Shigeno, Shuichi, et al. "Cephalopod Brains: An Overview of Current Knowledge to Facilitate Comparison with Vertebrates." *Frontiers in Physiology* 9, no. 952 (2018). DOI:10.3389/fphys.2018.00952.

Simpson, G. C. "The Baldwin Effect." *Evolution* 7 (1953): 110–17.

Singer, Emily. "Bat Echoes Used as Virtual Reality Guide." *New Scientist,* September 14, 2003.

Siviy, S. M. "Neurobiological Substrates of Play Behavior: Glimpses into the Structure and Function of Mammalian Playfulness." In *Animal Play: Evolutionary, Comparative and Ecological Perspectives,* edited by Mark Bekoff and John A. Byers, 221–42. Cambridge: Cambridge University Press, 1998.

Spinka, Marek, Ruth C. Newberry, and Marc Bekoff. "Mammalian Play: Training for the Unexpected." *Quarterly Review of Biology* 76 (2001): 141–68.

Stacho, Martin, et al. "A Cortex-Like Canonical Circuit in the Avian Forebrain." *Science* 369, no. 6511 (2020): eabc5534. DOI:10.1126/science.abc5534.

Sweis, Brian M. et al. "Mice Learn to Avoid Regret." *PLoS Biology* 16 (2018).

Takeshita, R. S. C., et al. "Beneficial Effect of Hot Spring Bathing on Stress Levels in Japanese Macaques." *Primates* 59 (2018): 215–25. DOI:10.1007/s10329-018-0655-x.

Thierry, Bernard. "Unity in Diversity: Lessons from Macaque Societies." *Evolutionary Anthropology: Issues* 16 (2007).

Thomas, E., and F. Schaller. "Das Spiel der Optisch Isolierten Kasper-Hauser-Katze." *Naturwissenschaften* 41 (1954): 557–58. Reprinted and translated in *Evolution of Play Behaviour*, edited by D. Muller-Schwarze. Stroudsburg, PA: Dowden, Hutchinson & Ross, 1978.

Thompson, Edward. *The Passions of Animals*. London: Chapman and Hall, 1851.

Thompson, Katerina V. "Self-Assessment in Juvenile Play." In *Animal Play: Evolutionary, Comparative and Ecological Perspectives*, edited by Mark Bekoff and John A. Byers. Cambridge: Cambridge University Press, 1998.

Thwaites, Thomas. *GoatMan: How I Took a Holiday from Being Human*. New York: Princeton Architectural Press, 2016.

Tinbergen, N. "On Aims and Methods of Ethology." *Zeitschrift für Tierpsychologie* 20 (1963): 410–33.

Tzar, Jennifer, and Eric Scigliano. "Through the Eye of an Octopus: An Exploration of the Brainpower of a Lowly Mollusk." *Discover*, January 19, 2003.

Ungurean, Gianina, et al. "Evolution and Plasticity of Sleep." *Current Opinion in Physiology* 15 (2020): 111–19.

Vandervert, Larry. "Vygotsky Meets Neuroscience: The Cerebellum and the Rise of Culture through Play." *American Journal of Play* 9 (2017): 202–27.

Warneken, Felix, and Alexandra G. Rosati. "Cognitive Capacities for Cooking in Chimpanzees." *Proceedings of the Royal Society B: Biological Sciences* 282 (2015).

Washburn, Margaret Floy. *The Animal Mind. A Textbook of Comparative Psychology*. 3rd ed. New York: Macmillan, 1926.

Watanabe, S., et al. "Pigeons' Discrimination of Paintings by Monet and Picasso." *Journal of the Experimental Analysis of Behavior* 63, no. 2 (1995): 165–74. DOI:10.1901/jeab.1995.63-165.

Waters, Dean A., and Husam H. Abulula. "The Virtual Bat: Echolocation in Virtual Reality." Semantic Scholar, 2001.

Watson, D. M. "Kangaroos at Play: Play Behaviour in the Macropodoidea." In *Animal Play: Evolutionary, Comparative and Ecological Perspectives*, edited by Mark Bekoff and John A. Byers. Cambridge: Cambridge University Press, 1998.

Watson, D. M., and D. B. Croft. "Age-Related Differences in Play-Fighting Strategies of Captive Male, Red-Necked Wallabies (*Macropus rufogriseus banksianus*)." *Ethology* 102 (1996): 336–46.

Weismann, August. *Das Keimplasma: Eine Theorie der Vererbung*. Jena, Germany: Fischer, 1892.

Whiten, Andrew. "The Burgeoning Reach of Animal Culture." *Science* 372 (2021).

Whiten, A., and R. W. Byrne. "Tactical Deception in Primates." *Behavioral and Brain Sciences* 11, no. 2 (1988): 233–44. DOI:10.1017/S0140525 X00049682.

Wilson, Edward O. *The Origins of Creativity*. New York: Liveright, 2017.

———. *Sociobiology*. Abr. ed. Cambridge, MA: Belknap Press of Harvard University Press, 1980.

Wilson, S. C., and D. G. Kleiman. "Eliciting Play: A Comparative Study." *American Zoologist* 14 (1974): 341–70.

Wozniak, R. H., and J. A. Santiago-Blay. "Trouble at Tyson Alley: James Mark Baldwin's Arrest in a Baltimore Bordello." *History of Psychology* 16, no. 4 (2013): 227–48. DOI:10.1037/a0033575. PMID:23914848.

Wu, Katherine J. "Zebra Finches Dream a Little Dream of Melody." *Smithsonian Magazine*, August 7, 2018.

Yeh, Pamela J., and Trevor D. Price. "Adaptive Phenotypic Plasticity and the Successful Colonization of a Novel Environment." *American Naturalist* 164, no. 4 (2004).

Yoshida, Ma., K. Yura, and A. Ogura. "Cephalopod Eye Evolution Was Modulated by the Acquisition of Pax-6 Splicing Variants." *Scientific Reports* 4, no. 4256 (2014). DOI:10.1038/srep04256.

Yuqian Ma, et al. "Mammalian Near-Infrared Image Vision through Injectable and Self-Powered Retinal Nanoantennae." *Cell* 177, no. 2 (2019): 243–55. DOI:10.1016/j.cell.2019.01.038.

Index

Ackerman, Diane, 116
adaptation, xiv
adaptive advantages, xiii, 13, 20,
    24, 27–30, 39, 45n, 54, 63,
    119, 125, 129, 145, 152,
    170, 181, 196, 210
  of dreams, 145, 147, 148
  eyes and, 180
  of play fighting, 70, 78, 161
  of running, 44
  of sleep, 145–47
  surplus resource theory and,
    156
  threats and, 71
  see also natural selection;
    reproduction; survival
adipose tissue, 29
adult stage, 57
Aelian (Claudius Aelianus), 5
affective neuroscience, 67
Ambrose, St., 211
American Psychological
    Association, 194, 224
amoebas, 224
amygdalae, 83–86, 92
Anderson, Roland, 3–7, 15, 169
animal, use of word, 3n

Animal Intelligence (Romanes),
    213–14
Animal Mind, The (Washburn), 224
animal minds and experiences,
    understanding, 209–14,
    224–25
  anecdotal evidence and, 218
  anthropomorphism and, see
    anthropomorphism
  assistive technologies and, 223,
    224
  Eiseley's experience and, 219
  Foster's attempt at, 227–30
  literary attempts at, 223–24
  sensory information and,
    221–23
  Thwaites' attempt at, 225–29
  umwelten and, 220–21, 224
animal play
  comparing across species and
    classes, 9
  defining and characterizing,
    x–xiii, 9–16, 24; see also
    Burghardt's definition of play
  difficulty of observing, x, 33
  lack of scientific interest in,
    x–xii

Animal Play Behavior (Fagen), xi,
    10n, 24
animals, humans becoming,
    207–9
animals becoming human, 209
animal species, number of, 164
anthropomorphism, 214–16, 218
    critical, 219, 220, 224
    by omission, 221
ants, 173, 180
Ants, The (Hölldobler and
    Wilson), 173
Aquarium of the Pacific, 171
Aristotle, 211, 245n
Atlantic, 120
Augustine, St., 211
Australian magpies, 122, 124–25,
    136
Avian Brain Nomenclature
    Consortium, 245n

baboons, 20
badgers, 227–28
balance, recovering, 45–48, 50, 76
Baldwin, James Mark, 194–99,
    202, 250n, 251n
Baldwin effect, 199–203
Barber, Nigel, 28–30, 33
basal ganglia, 116
baseball, 235–36
Bateson, Sir Patrick, 126
bats, 223, 224–25
Beagle, HMS, 43, 100
bears, 23, 31–32
beauty, 135–36
bees, 173–74, 222
behavioral flexibility, 165, 167,
    177, 178, 197, 198, 200
behavioral neuroscience, 64, 67
Being a Beast (Foster), 227

Being and Time (Heidegger), 226
Bekoff, Marc, 25, 44–45, 47, 52,
    63, 76, 98, 110, 202
Belding's ground squirrels, 31
Bell, Heather, 82
Biben, Maxeen, 49–50, 75
Bible, 251n
birds, 5, 29, 136, 166, 167, 218
    aerial food passes between,
        123, 136
    bowerbirds, 135
    brain of, 116–17
    calls of, 124
    crows, 8, 117, 118, 120–21,
        219
    energy expenditure of, 171
    herring gulls, 120–21, 136
    intelligence of, 116–18
    juncos, 199–200
    keas, 121–22, 246n
    magpies, 122, 124–25, 136
    migration of, 230
    Montagu's harriers, 122–23,
        125, 136
    object play in, 120, 121, 136
    parrots, 117, 118, 135n, 144
    peacocks, 135
    pigeons, 117–18
    play fighting in, 122, 124–25,
        136
    ravens, xii, 115, 119, 120,
        126–27, 129
    singing of, 124, 125, 150
    sleep and dreaming in, 143–46,
        150
    soaring of, 119n
    varieties of play in, 118–19
    zebra finches, 150
Blake, William, 221
Boal, Jean, 5, 15

Bond, Alan B., 121–22
bones, 155
bonobos, 59
boredom, 15
bowerbirds, 135
Bowling Green State University, 67
brain, xii, xiii, 11, 25, 46, 51, 63,
        78–80, 92, 173, 175, 221
    amygdalae in, 83–86, 92
    basal ganglia in, 116
    of birds, 116–17
    brain stem in, 79, 80, 86n
    cerebellum in, 79–80, 86n
    cerebrum in, 79, 80, 83, 86n
    cortex in, 80–82, 245n
    dreaming and, 144
    evolution of, 245n
    imaging techniques and, xii
    innovation and, 185–87
    maturing of, 27
    neocortex in, 116, 245n
    neurons in, see neurons
    prefrontal cortex in, 81–85, 92
    of rats, 79–85
    sleep and, 146, 147
    synapses in, 26, 83, 141
    of vertebrates, 173
Brown, Stuart, 87
Browning, Elizabeth Barrett, 224
Brownlee, Alex, 25, 43
Buddhism, 217, 237
Bulkeley, Kelly, 140–41, 145, 147,
        151
Burghardt, Gordon, xi, 12–13,
        163, 168–70, 210–11,
        218–19, 221
    on conditions that enable play,
        164–67, 170, 177
    surplus resource theory of,
        90–91, 155–56, 178

Burghardt's definition of play,
        13–15, 59, 121, 122, 124,
        130
    applied to bees' behavior, 174
    applied to dreaming, 141–42
    as characterized by repeated
        but varied movements,
        13–15, 130, 141–42, 168,
        174
    as nonfunctional, 13, 14, 130,
        141, 142, 168, 174
    as performed by well-fed, safe,
        and healthy animals, 14,
        130, 142, 168, 174
    Thwaites' and Foster's
        experiments and, 229
    as voluntary, 13, 14, 130, 141,
        142, 168, 174
Byers, John, 25–26, 76n, 83

calories, 165
Cambrian explosion, 175, 176
Cambridge Declaration on
        Consciousness, 241n
Cameron, Elissa, 31, 32
Campbell, Keith, 43
capuchins, 128, 129, 134, 156,
        178, 196–98
Caro, Tim, 27–28
cats, 26
    playing with mice, 27–28
    scent of, and rats, 71
cause, pleasure of being, 127, 128
Centers for Disease Control, 79n
cephalopods, 5
cerebellum, 79–80, 86n
cerebrum, 79, 80, 83, 86n
chamois, 55–56
chemistry, 58
child development, 194

child-rearing, 165–67, 186–87, 200

chimpanzees, 27, 59, 62, 96, 134, 144, 157
  sweet potato slices and, 132–33

Chordata, 5

chromosomes, 179

circadian rhythms, 147

cladistics, 158

cladogram, 158, 159, 161, 177, 181

Clever Hans, 214–15, 221

closed systems, 92

cognitive ethology, 217

cognitive neuroscience, 64, 67

competition and collaboration, 99–104

conflict, defusing, 91, 104
  in rats, 73–74, 78, 103

consciousness, 141, 145, 216, 241n

contagion effect, 53, 83

control, losing and regaining
  recovering balance, 45–48, 50, 76
  relief and, 109

Cope, Edward Drinker, 191

cortex, 80–82, 245n

corvids, 118–19
  crows, 8, 117, 118, 120–21, 219
  ravens, xii, 115, 119, 120, 126–27, 129

creativity, 149–52

Cristol, Daniel, 120–21

crows, 8, 117, 118, 120–21, 219

culture
  play and, 124–25, 129–30
  refinements in, 134
  self-handicapping in, 125
  Whiten's definition of, 125, 130

culture, animal, xiii
  development of, 134–35, 136
  play and, xii, 135

cuttlefish, 139, 144, 145

dangerous behaviors
  in juvenile humans, 55
  in monkeys, 53–55

dark-eyed juncos, 199–200

Darwin, Charles, xiv, 22, 28n, 30, 39, 43, 99–101, 102n, 135n, 188–91, 195, 212, 213, 233, 249n
  on birds, 119n
  children of, 56, 76–77, 155
  The Descent of Man, 100n, 189, 212, 250n
  on dreaming, 143
  The Expression of the Emotions in Man and Animals, 212
  The Formation of Vegetable Mould through the Action of Worms, 212
  on insects, 248n
  on language, 152
  on monkey intelligence, 127
  nature as viewed by, 193
  On the Origin of Species, xivn, xv, 58n, 101, 152, 155, 178–79, 188–91, 193, 249n
  religious sentiments of, 250n
  sexual selection theory of, 135–36
  on Tree of Life, 58

Darwinists, 188, 190, 191, 193, 198–99

deception, 103
  in play fighting, 103–8

Descent of Man, The (Darwin), 100n, 189, 212, 250n

degus, 99, 160–61, 185
Dennett, Daniel, 200
Descartes, René, 211–12
determinism, 193
developmental stages, 51
    and training for the
        unexpected, 56–58
de Waal, Frans, 217
Dewey, John, 233
Diamond, Judy, 121–22
diet, *see* food
dinosaurs, 167
Djungarian hamsters, 159–60
DNA, 157
dogs, 21, 27, 49, 151, 170, 202,
    222, 237
    dreaming in, 143
    gazing behavior in, 97, 111
    horses playing with, 230–31
    humans' relationship with,
        95–96, 97
    morality in, 110–11
    play bow of, 97–98, 99
    play fighting in, 98–99
    self-handicapping in, 49, 52
    sliding activities in, 55
    snow and, 51–53, 55, 229
    wolves' evolution to, 95–97
dolphins, xii, 130–31
dopamine, 80
dorsal ventricular ridge, 116
dreaming, 139–40
    adaptive advantage of, 145,
        147, 148
    behavioral manifestations of,
        139, 142–43
    in birds, 143–45, 150
    brain and, 144
    Burghardt's definition of play
        applied to, 141–42

creativity and, 149–52
    in cuttlefish, 139
    difficulties of defining and
        studying, 140
    electroencephalography in
        study of, 144
    nightmares in, 142, 148
    in octopuses, 145
    play theory of, 140–41, 145,
        148, 151
    recurring elements in, 142
    self-handicapping in, 149
    sleep-talking and, 143–45
    social simulation theory of, 148
    threat simulation theory of, 148
    as training for the unexpected,
        148–49
    waking dreams, 151
Duke University, 96

eagles, 119n
ectotherms, 167, 169, 170
Ediacaran period, 175
Edinburgh, 43
Edinburgh Pig Park, 43–44
Edinger, Ludwig, 116, 117, 245n
effect, pleasure of having, 127,
    128
Eimer, Theodor, 191
Einon, Dorothy, 71–72, 78
Eiseley, Loren, 219
Eldridge, Niles, 250n
electroencephalography, 144, 147
elephants, xii, 222
    sliding activities in, 56
Eliade, Mircea, 235, 236
elk, 23
embryology, 179
emotions, 10, 20–21, 46, 67, 83,
    90, 218, 224

empathy, 90
endosymbiosis, 103
endotherms, 165–67, 169
energy, 19, 39, 49, 63
    required to move, 170–71
    surplus, 21, 29, 90, 141,
        155–56, 164
Eno, Brian, 125
entropy, 92
equifinality, 126
estimativa, 211
ethogram, 58–62
ethologist, defined, xiii
eukaryotic cells, 103
evolution, xiii–xv, 9, 39, 44, 63,
        78
    of brain, 245n
    controversies about, 187–90
    convergent, 180
    fossil record and, 191
    Lamarckian, 188, 190–93, 195,
        196, 198, 233
    of languages, 152
    Modern Synthesis, 249n
    natural selection in, *see* natural
        selection
    organic selection in, 194–99,
        250n, 251n
    and physical similarities
        between species, 157–58
    of play, 152, 155–62, 175–76,
        178, 180–81
    self-directed, 187, 198–201,
        203
    sexual selection in, 135–36
    and shared characteristics of
        species, 158
    success in, 34
    time needed for, 249n
Evolution, 199

evolutionary biology, 157–58
    developmental, 179
    phylogenetic categories in, 58,
        59n, 157, 158, 167, 170
    radiations in, 161, 177
exaptation, 247n
excitement, 47
exercise, 10, 26
exploration and investigation,
        12–14, 48, 51
*Expression of the Emotions in Man and
        Animals, The* (Darwin), 212
eyes, 179, 180

Fagen, Johanna, 31–32
Fagen, Robert, xi, 10n, 23–24,
        31–33
Fairbanks, Lynn, 26–27
fairness, 98–99, 104, 110
falls, recovering from, 45–48, 50
fear, 46, 84
fear circuit, 84
Ficken, Millicent, 118–19
fighting, 91
    in keas, 122
    in meerkats, 36, 50
    in rats, 69, 74, 82, 105
    see also play fighting
fight-or-flight response, 86
finches, 150
fire, 132
fireflies, 103
fish, 147
    energy expenditure of, 171
    leaping of, 172
    play in, 170–73
food
    birds' passing to partners, 123,
        136
    cooked, 131–33

dietary requirements for play, 165–67
foraging for, 178, 202
Formation of Vegetable Mould through the Action of Worms, The (Darwin), 212
Foster, Charles, 227–30
foxes, 27, 229–30
free will, 193–94, 195, 198
fun, 9, 47–48, 52, 68
fungi, 157

Gamble, Jennifer, 120–21
game face, 12
geese, 100–101
genes, 178–80
    in evolution of play, 180–81
    master, 179–81
    Mendelian genetics, 179, 199, 249n
Genesis of Animal Play, The (Burghardt), 13
giraffes, 191
God, 192–93, 211, 250n
Godfrey-Smith, Peter, 139
Goodfellas, 108–9
gorillas, 185–87
Gould, Stephen Jay, 247n, 250n
Gray, Asa, 188
Gray, Peter, 88
Greyfriars Bobby, 43
Griffin, Donald, 216–17
Groos, Karl, 27, 29, 52, 55–56, 116, 121, 127, 128, 130–31, 202, 216, 242n
    The Play of Animals, 21–22, 27, 151, 190, 194, 195, 250n
    practice hypothesis of, 21–22, 60, 148, 194; *see also* practice hypothesis

Hall, G. Stanley, 242n
Hall, Sarah L., 28
hamsters, 159–60
Harcourt, Robert, 20
Hare, Brian, 96, 97
Haslam, Michael, 134
hawks, 119n, 180
Heidegger, Martin, 226
Heinrich, Bernd, 115, 120
Henslow, George, 191
herring gulls, 120–21, 136
hibernation, 29
Hölldobler, Bert, 173
Homo Ludens (Huizinga), 97–98
Hooker, Joseph, 249n
Horowitz, Alexandra, 224
horses, 31, 146, 167
    Clever Hans, 214–15, 221
    dogs playing with, 230–31
Huber, Pierre, 248n
Huizinga, Johan, 97–98, 118, 129, 235, 236
human mating behavior, 163
human play in adults, x, 12, 27, 242n
    negotiation, assessment, and manipulation in, 108–10
human play in children, x, 8, 22, 56, 126, 242n
    decline of, 87–88
    play fighting, 89–90
    rough-and-tumble, 88–90
humans, animals becoming, 209
humans taking animal forms, 207–9
Humboldt, Alexander von, 135n
Hume, David, 211–12
hunting, 22, 178, 202
    cats and, 27–28
    raptors and, 119n
    wolves and, 23

Huxley, Julian, 199
Huxley, Thomas Henry, 188
Hyland, Drew, xi
hypotheses, 24

ibex, 19, 51
Imanishi, Kinji, 217
immune system, 71
innovation, 185–87, 202–3, 218
insects, 136, 147, 173–75, 248n
    bees, 173–74, 222
Inside of a Dog (Horowitz), 224
instincts, 22, 211
intelligence, 213–14, 221
    of birds, 116–18
    of monkeys, 127–28
    play and, 116, 118
intelligent design, 28n
intention, 10, 14
interspecies play, 172, 230–31
investigation and exploration,
    12–14, 48, 51

jellyfish, 147
Johns Hopkins University, 199
Johnson, Steven, 129–30
juncos, 199–200
Just So Stories (Kipling), 32–33
juvenile period, 57
    dangerous behaviors in, 55
    extended, 165–67, 177, 187

Kalahari Desert, 19, 36
    meerkats in, see meerkats
Kalsoy, 207–9
kangaroos, 29, 49, 164, 176, 164
keas, 121–22, 246n
Kendall, May, 225n
Kipling, Rudyard, 32–33
Komodo monitor lizards, 169–70

Kropotkin, Pyotr, 101–2
Kubrick, Stanley, 129

Laban movement analysis, 74–75
Lamarck, Jean-Baptiste, 190–91,
    245n
Lamarckians, 188, 190–93, 195,
    196, 198, 233
language, 118, 135n, 152, 221, 237
langurs, 59–62
Lefebvre, Louis, 218
lemurs, 49, 53, 59, 162
lizards, 167, 169–70
lorises, 8, 186–87
Lyell, Charles, xivn, 188

macaques, 59, 105–8, 134, 162,
    186
Macquarie University, 20
magic circle, 235
magic tricks, 110
magnetic resonance imaging, xii
magpies, 122, 124–25, 136
mammals, 5, 67, 164, 166, 170, 186
    hibernating, 29
    juvenile period of, 22
    marsupials, 163–64
    monotremes, 248n
    placental, 163–64
    sliding activities in, 55–56
Marek, Roger, 84
Margulis, Lynn, 102–3
marsupials, 163–64
Mather, Jennifer, 3–8, 15–16, 169,
    178
McCartney, Paul, 150
McKenna, Mario, 86n
meerkats, 19, 32–39, 53
    fighting and play fighting in,
    34–36, 50

practice hypothesis tested with,
34–36, 38
social bonding hypothesis
tested with, 36–39, 72
survival of, 36–37
Mei Xiang, ix
memory, 117, 146
Mendel, Gregor, 179
Mendelian genetics, 179, 199,
249n
mental illness, 67
metabolic rate, 164–67, 169–71,
177
methodological behaviorism,
216–17
mice, 11, 26, 79n
cats' playing with, 27–28
culinary skills of, 131–33
energy expenditure of, 171
play fighting in, 158–59, 166
Midsummer Night's Dream, A
(Shakespeare), 223
Mikladalur, 207–8
mind, 211
mitochondria, 102
molecular genetics, 179
moles, 222
mollusks, 144, 176
cuttlefish, 139, 144, 145
octopuses, *see* octopuses
Monash University, 33
mongooses, 98
monitor lizards, 169–70
monkeys, 19–20, 26–27, 49, 146,
162, 186, 202
belly-flopping behavior in,
53–55
capuchin, 128, 129, 134, 156,
178, 196–98
innovation in, 185

macaques, 59, 105–8, 134, 162,
186
parkour behavior in, 62, 185,
229
Romanes' study of, 127–29
self-handicapping in, 59–62,
152
somersaulting behavior in,
59–62
monotremes, 248n
Montagu's harriers, 122–23, 125,
136
moose, 23
moral behavior, 110–11
Morgan, C. Lloyd, 195, 215–16
Morgan, Michael J., 71–72
motive, 10
mouth, relaxed expression of,
230–31
murderers, 87
murids, 158–61
muscles and motor functions, 25,
27, 46, 76, 156, 243n
musicality, 201
mutationists, 188, 192
Mutual Aid (Kropotkin), 101
Myllokunmingia, 176

Nagahama Institute of Bio-Science
and Technology, 180
Nagel, Thomas, 224–25
National Institute for Play, 87
National Zoo, 168–69, 171
natural selection, xiii, 30, 45, 47,
135, 189–91, 229, 234, 245n
Baldwin effect and, 199–203
costs of, 39
critiques of, 188, 249n
eyes and, 179, 180
fitness in, 189

natural selection (*cont.*)
  free will and, 193–94, 195
  frugality of, 155
  genes and, 178–79
  inherited characteristics in, 189
  play as influence on, 185–203
  play influenced by, 155–81
  resource competition in, 189
  sleep and, 145–46
  surplus resource theory and, 156
  survival of the fittest in, xivn
  see also adaptive advantages
natural selection characteristics, and characteristics of play, xiv–xv, 16, 64, 234
  beauty and, 135–36
  competition and collaboration and, 99–104
  intangible realms and, 152
  order and, 91–92
  preparation for the unexpected and, 63
  time and energy costs and, 39
nature, 192–94, 233–34
nature-deficit disorder, 88
neocortex, 116, 245n
neo-Lamarckians, 188, 190–93, 195, 196, 198, 233
nervous system, xiii, 11, 26, 27, 46, 51, 63, 78, 80, 91, 92, 173, 175
  of rats, 79
  sleep and, 145, 146, 147
neurons, 80, 84
  circuits of, 178, 197, 201
  of jellyfish, 147
  networks of, xii, 79, 84, 92
  synapses and, 26, 83, 141
neuroscience, xii, 64, 67, 86n

neurotransmitters, 80, 116
Newberry, Ruth, 44–45, 47, 52, 53, 76, 202
newborn stage, 57
Newton, Isaac, 126
Nunes, Scott, 31, 32

object play, 8, 13, 48, 168–69
  in birds, 120, 121, 136
  developmental stages and, 57
  in octopuses, 3, 5–8, 14–16, 169, 173, 177, 178
  self-handicapping in, 49, 50, 52
octopuses, 3–8, 14–16, 144, 177–78, 180
  arms of, 3n
  dreaming in, 145
  exhalant funnels of, 3, 7, 8, 15, 16
  eyes of, 180
  foresight and planning in, 4
  inhumane research on, 241n
  intelligence and curiosity of, 5–6
  play in, 3, 5–8, 14–16, 169, 173, 177, 178
  senses of, 7
On the Origin of Species (Darwin), xivn, xv, 58n, 101, 152, 155, 178–79, 188–91, 193, 249n
orangutans, 110, 157
orchids, 136
order, 91–92
organelles, 102–3
organic selection, 194–99, 250n, 251n
orthogeneticists, 188, 193
Osborn, Henry Fairfield, 195, 251n
oxytocin, 97, 111

Palagi, Elisabetta, 230–31
Panda Cams, ix
Pandolfi, Massimo, 123
Panksepp, Jaak, xi, 67–68, 79,
    88–89, 150, 222, 223
parenting, 165–67, 186–87, 200
parrots, 117, 118, 135n, 144
    keas, 121–22, 246n
Passions of Animals, The
    (Thompson), 20–21, 143
PAX6 gene, 180
peacocks, 135
Pellis, Sergio and Vivien, 53–54,
    74–75, 77–81, 84, 86,
    86n, 87, 89–92, 105, 122,
    156–59, 162, 185, 186, 210
Pepperberg, Irene, 118
Petrů, Milada, 59–61, 63, 185, 202
Pfungst, Oskar, 215
phenotypes, 196
Philosophie Zoologique (Lamarck),
    190
phylogenetic tree, 58, 59n, 157,
    158, 167, 170
Piaget, Jean, 194
pigeons, 117–18
pigs, 43–44, 53
    flop-over behaviors in, 44–48, 76
pit vipers, 222, 223
placental mammals, 163–64
play, definitions of, x, 9, 235,
    242n
play face, 98
play fighting, 22, 50, 53, 89, 185
    adaptive advantages of, 70, 78,
        161
    in capuchins, 156
    in children, 89–90
    deception and ambiguity in,
        103–8
    in degus, 99, 160–61, 185
    in dogs, 98–99
    fairness in, 98–99, 104, 110
    in gorillas, 185
    in kangaroos, 164
    in magpies, 122, 124–25, 136
    in meerkats, 34–36, 50
    in mice, 158–59, 166
    in murids, 158–61
    in primates, 162
    proactive skills developed in,
        103–4
    in rats, 69–78, 82, 84–86, 99,
        103, 108, 157, 158, 160,
        166, 185
    in rats, moves employed in,
        75–76, 81, 82, 86n, 158, 160
    self-handicapping in, 49–50
    sex and, 70, 74, 78, 160–63
    and social competency, in
        rodents, 71–72, 77, 78, 82,
        103, 161
    as training for the unexpected,
        90, 161
Playful Brain, The (Pellis and
    Pellis), 87
Play of Animals, The (Groos), 21–22,
    27, 151, 190, 194, 195, 250n
play patterns, 60–61
pleasure, 9, 12, 27, 119
Pliocene epoch, 37
Podlipniak, Piotr, 201
pointing cues, 96
positron-emission tomography, xii
pottos, 162–63
practice hypothesis, 21–22, 29, 33,
    34, 57–58, 60, 194
    and cats' playing with mice,
        27–28
    meerkats and, 34–36, 38

practice hypothesis (*cont.*)
  motor training and, 25, 27
  threat simulation theory and, 148
predators and prey, 22, 44–46, 49,
    50, 63, 135, 145, 172, 178,
    180, 231
prefrontal cortex, 81–85, 92
Preyer, W., 126–27
Price, Trevor, 200
primary process play, 90–91,
    155–56, 165, 176
primates, 27, 29, 49, 59, 62, 98,
    117, 162, 186
  chimpanzees, *see* chimpanzees
  cultural differences among, 217
  innovation in, 185–87
  lemurs, 49, 53, 59, 162
  lorises, 8, 186–87
  monkeys, *see* monkeys
  play fighting in, 162
  pottos, 162–63
  tools used by, 134
Princeton University, 194
Principles of Geology (Lyell), 188
prokaryotic cells, 103
prosocial behaviors, 96
pro wrestling, 151
Psychological Review, 194
psychopathology, 87–88

Queen Mary University of
    London, 174
Question of Animal Awareness, The
    (Griffin), 216–17
Question of Play, The (Hyland), xi
quiescence, 147

rabbits, 196–98
radiations, in evolutionary biology,
    161, 177

raptors, 119n
  eagles, 119n
  hawks, 119n, 180
  Montagu's harriers, 122–23,
    125, 136
rats, 11, 26, 49, 79, 222
  brains of, 79–85
  chirping of, 68, 150
  conflict de-escalation in, 73–74,
    78, 103
  fighting in, 69, 74, 82, 105
  nervous systems of, 79
  nuzzling in, 69, 70, 77
  play fighting in, 69–78, 82,
    84–86, 99, 103, 108, 157,
    158, 160, 166
  play-fighting moves of, 75–76,
    81, 82, 158, 160
  practice for sex in, 70, 74, 78
  social competency in, 71–72,
    77, 78, 82, 103
  social status among, 73–74,
    104–5
  stress reduction in, 71, 78
  tickling of, 68, 76
ravens, xii, 115, 119, 120, 126–27,
    129
reason, 211, 212, 237
recapitulation theory, 242n
relaxed open-mouth display,
    230–31
relief, 109
religion, 233, 236–37, 250n, 251n
  God, 192–93, 211, 250n
reproduction, xiii, xv, 13, 14, 22,
    30, 31, 34, 70, 119, 125,
    141, 147, 189
reproductive cells, 192
reptiles, 147
  lizards, 167, 169–70

recreation in, 166–69
turtles, 166–70, 171
Research Institute of Animal
     Production, 44
Revonsuo, Antti, 148
Ritter, Erich, 172–73
rodents
     degus, 99, 160–61
     guinea pig–like, 160, 161
     mouse-like, 160
     murids, 158–61
     squirrels and squirrel-like, 31,
          160, 161
     see also mice; rats
Romanes, Charlotte, 128–29
Romanes, George, 43, 127–28,
     143, 144, 212–14
Rosati, Alexandra G., 132–33
Roslin Institute, 43
running, 44, 45

Sacred and the Profane, The (Eliade),
     235, 236
satisfaction, 9
Schiller, Friedrich, 21
science, 233
     basic versus applied, 125–26
     theories and hypotheses in, 24
scientific research on animal
     behavior
     environmental settings for, 33
     methodological behaviorism
          and, 216–17
     rigor needed in, 214–16
scientific research on animals
     for drug development, 79n
     humane concerns and, 11, 79n,
          144, 241n
     on play, lack of interest in, x–xii
Scorsese, Martin, 108

sea lions, 20, 53
seals, 20, 52, 202
Seattle Aquarium, 3, 4
secondary process play, 91,
     155–56, 176, 178
self-handicapping, 48–52, 59, 60,
     202
     culture and, 125
     in dreams, 149
     in interspecies play, 230, 231
     in monkeys, 59–62, 152
     in Thwaites' and Foster's
          experiments, 229
selkies, 207–9
Senckenberg Research Institute
     and Natural History
     Museum, 172
senses, 221–23
sensitive period hypothesis, 26, 76n
serotonin, 80
sex, 91
     play fighting and, 70, 74, 78,
          160–63
     see also reproduction
sexual selection, 135–36
sharks, 172–73, 222
Sharpe, Lynda, 19, 20, 32–36,
     38–39, 50, 53, 63, 72, 78,
     81n, 88
sheep, 43, 51, 211
sight, 222, 223
sleep, 145
     adaptive advantages of, 145–47
     in birds, 143–46, 150
     brain and, 146, 147
     NREM, 146–47
     REM, 146–47, 150
     talking during, 143–45
     varied forms of, 146–47
     see also dreaming

Sleep and Dream Database, 140
sleep signatures, 147
sliding, 55–56, 168
    snowboarding crow, 120–21
smell, sense of, 222
Smithsonian National Zoo, ix
Smolker, Rachel, 115, 120
snow, 51–53, 55–56, 229
snowboarding crow, 120–21
social bonding, 21–23, 33, 34, 72,
    148
    dreams and, 148
    meerkats and, 36–39, 72
social competency, 72–73
    and play fighting, in rodents,
        71–72, 77, 78, 82, 103, 161
social play, 8, 13, 31, 169
    developmental stages and, 57
    between fish and human, 172
    self-handicapping in, 49
    see also play fighting
social simulation theory of
    dreams, 148
solitary play, 8, 13, 167, 168, 171
    developmental stages and, 57
    self-handicapping in, 49, 50, 52
somatic cells, 192
somniloquy, 143
soul, 211, 212, 217, 251
Spencer, Herbert, xivn, 21, 29, 90,
    141
Spinka, Marek, 44–45, 47–50, 52,
    53, 57, 59, 63, 76, 110, 146,
    148–49, 202, 229
squid, 144
squirrels and squirrel-like rodents,
    31, 160, 161
star-nosed moles, 222
Stellenbosch University, 33
stereotyped behaviors, 11–14, 91

stress, 71, 78
surplus resource theory, 90–91,
    155–56, 178
survival, xiii, 13, 14, 22, 30, 31,
    34, 71, 119, 125, 141, 145,
    147, 242n
    of bears, 31–32
    of the fittest, xivn
    of meerkats, 36–37
    play as hindrance to, 19–20
symbiosis, 102–3
synapses, 26, 83, 141
Syrian golden hamsters, 159

tertiary process play, 91, 155–56,
    176
theories and hypotheses, 24
theory of mind, 98, 104, 227
therianthropes, 209
thermogenesis, 29
Thierry, Bernard, 106–7
Thomas Aquinas, St., 211
Thompson, Edward, 20–21, 143,
    212
Thompson, Katerina V., 29, 30,
    33, 50
threat, 71, 165
threat simulation theory, 148
Thwaites, Thomas, 225–29
Tian Tian, ix
tickling, 68, 76, 77
Tinbergen, Nikolaas, 78, 152, 210
tinkering, 126
Tolstoy, Leo, 223–24
Tomasello, Michael, 96
tools, 129, 134
touch, sense of, 222
training for the unexpected,
    45–48, 50, 59, 63, 202
    developmental stages and, 56–58

dreams as, 148–49
in langurs, 59–62
magic tricks and, 110
in monkeys, 54–55
play fighting as, 90, 161
predictions of hypothesis of, 51–53
in rats, 76–78
tickling and, 77
transitive inferential logic, 117–18
Tree of Life, 58
turtles, 166–70, 171
2001: A Space Odyssey, 129

umwelten, 220–21, 224
Université Louis-Pasteur, 106
University College London, 71
University of Alaska, Fairbanks, 23
University of Berlin, 215
University of California, Los Angeles, 26
University of California, San Diego, 200
University of Cambridge, 126
University of Colorado, Boulder, 25, 44
University of Edinburgh, 43
University of Idaho, 25
University of Lethbridge, 53
University of Oxford, 134
University of Pisa, 230
University of St. Andrews, 27
University of Southampton, 28
University of Tennessee, Knoxville, 13

University of Toronto, 194
University of Urbino, 123
Unleashed (Hempel and Shepard, eds.), 224

vaccines, 79n
verbal play, 108
vertebrate brains, 173
voles, 98, 159
von Uexküll, Jakob, 220–21
Vrba, Elisabeth S., 247n

Walker, Curt, 26, 76n
Walker, Matthew, 150
Wallace, Alfred Russel, 188
Warneken, Felix, 132–33
Washburn, Margaret Floy, 224
Washington State University, 44
Weismann, August, 192
whales, 134
Whishaw, Ian, 80
Whiten, Andrew, 125, 130
Whitman, Charles, 87
Wilmut, Ian, 43
Wilson, Edward O., xi–xii, 173
Woolf, Virginia, 224
wolves, 23, 49
evolution to dogs, 95–97
Wonderland (Johnson), 129–30
Wood-Gush, David, 43–44
wrestlers, 151
wulst, 116

Yeh, Pamela, 200

zebra finches, 150